Railway Security

This book provides an overview and assessment of the security risks, both man-made and natural, facing the railways and rail networks.

Railroads face significant threats from disasters, but with situational awareness and coordinated effort these can often be substantially minimized. Transportation assets have always been vulnerable to natural disasters, but in the current environment these assets are also a preferred target of human-caused disruption, especially in the form of terrorism, as the events in many other parts of the world have underscored. Railways are not a homogeneous mode of transportation given their various roles in intercity and commuter passenger movement, as well as being a major portion of the freight ton-miles upon which the U.S. economy is highly dependent. Designed to provide advice for railway owners and first responders, this text discusses how to secure hazardous material transport and how to establish guidelines for rail freight operations and rail passenger operations. The book aims to develop an understanding of the unique operating characteristics of railways, the nature and the range of vulnerabilities, the present means for protecting the infrastructure, and the required public policy initiatives that are prerequisite for developing a comprehensive appreciation of the magnitude of this issue. The book utilizes case studies of transport disasters to illustrate lessons learned and to provide critical insight into preventative measures.

This book will be of great interest to students and practitioners of transportation, technology and engineering, and security management.

Richard R. Young is Professor of Supply Chain Management at the School of Business Administration, Pennsylvania State University, Harrisburg, USA, and a Fellow of the Chartered Institute of Logistics and Transport.

Gary A. Gordon (Lt. Col., USA, Ret.) is Adjunct Faculty at the University of Massachusetts Lowell, USA, and former Assistant Federal Security Director for DHS/TSA, Railroad Specialist for the former ICC, and Assistant Chief Engineer of a Northeast railroad.

Jeremy F. Plant is Professor Emeritus of Public Policy and Administration, Pennsylvania State University, Harrisburg, USA.

Railway Security

Protecting Against Manmade and Natural Disasters

Richard R. Young, Gary A. Gordon, and Jeremy F. Plant

Routledge
Taylor & Francis Group
LONDON AND NEW YORK

First published 2018
by Routledge
2 Park Square, Milton Park, Abingdon, Oxon OX14 4RN

and by Routledge
711 Third Avenue, New York, NY 10017

Routledge is an imprint of the Taylor & Francis Group, an informa business

British Library Cataloguing-in-Publication Data
A catalogue record for this book is available from the British Library

Library of Congress Cataloging-in-Publication Data
Names: Young, Richard R., 1946– author. | Gordon, Gary A., author. | Plant, Jeremy F., 1945– author.
Title: Railway security : protecting against manmade and natural disasters / Richard R. Young, Gary A. Gordon, and Jeremy F. Plant.
Description: Abington, Oxon ; New York, NY : Routledge, 2017. | Includes bibliographical references and index.
Identifiers: LCCN 2017006850 | ISBN 9781420080643 (hardback : alk. paper) | ISBN 9781315155296 (ebook)
Subjects: LCSH: Railroad accidents—Prevention. | Railroads—Security measures. | Terrorism—Prevention. | Emergency management.
Classification: LCC HE1779 .Y66 2017 | DDC 363.12/2—dc23
LC record available at https://lccn.loc.gov/2017006850

ISBN: 978-1-4200-8064-3 (hbk)
ISBN: 978-1-315-15529-6 (ebk)

Typeset in Times New Roman
by Apex CoVantage, LLC

We gratefully dedicate this book to our wives – Mary, Bobbie, and Susan – for the patience that they had for us during the long period of research and writing that went into this volume. They left us to the solitude of our respective studies and endured overhearing many long conference calls. We could not have done it without their encouragement and support.

Contents

Figures

Tables

Foreword

In our days together standing up the new Department of Homeland Security (DHS), Secretary Tom Ridge and I spoke often about our concern that understandable attention to commercial aviation security could be crowding out attention deserved by the other modes of transportation, including railways. The years following 9/11/01 held significant events in those other modes. Spain was attacked in Madrid in 2004. The target? Commuter rail. The United Kingdom was attacked in London in 2005. The target? The underground transit system. India was attacked in Mumbai in 2008. The mode of attack? From the sea. Other tragedies without a terrorism nexus abound. Deepwater Horizon, Hurricane Katrina, and Super-Storm Sandy come to mind here at home with sadly so many others around the globe. One of the few common elements to all these events is the impact each had on the multi-modal transportation system. . . whether or not that system was the focused target. The harsh reality in our ever-smaller globe is that any significant negative impact on even one mode can have ramifications locally, nationally, and even internationally. Think about simple things. . . that shirt you just bought at Walmart that was manufactured in China enjoyed an ocean voyage to the Port of Los Angeles, a train ride to a distribution warehouse, a truck ride to the retail store, and a ride in your car to its new home in your closet. In this new post-9/11 world we are still trying to understand, and during these days when change is a constant and the acceleration of the pace of that change demands ever-growing attention, security seems to have become the new dimension of everything. From the elements of our personal lives in this New Normal world to the massively complex challenges at the national security level, virtually every place we go, events we attend, choices of travel we make, and so many other daily decision elements are being refitted for their security realities. This excellent book focuses on one of those elements. . . security on railways. I wish I had been able to read it as part of our thinking in 2002 as we stood up DHS. In the same manner we have come to understand the linkages and overlaps among the multiple modes of transportation, we should recognize and learn from the multiple dimensions of any one of those modes. Just read the table of contents here to take an initial inventory of railway issues with security requirements. . . infrastructure, policing, protection, hazardous materials transport, emergency response, and on and on. The three authors recognize what I have come to believe is the most important word in the post-9/11 security vocabulary. . . *collaboration.* Each of them brings credible expertise to the book. An academic with a personal love of railroads and years of work looking through an academic public policy lens. A practitioner whose daily life depended on railways to get his products from Point A to Point B. His work mandated expertise in the supply chain and included the interface between maritime and rail transport. A transportation engineer with not only expertise in rail infrastructure but also real time in the Transportation Security Administration working rail security issues. Their collaboration offers the chance to see the rail security challenge from all sides, and perhaps most importantly, as the title of the final chapter suggests, this collaboration has literally just begun and the analytic work and conclusions reached are all just points on a moving timeline and the idea of final decisions is itself forever transient.

A recent article by Loren Thompson in *The National Interest* focuses on a subset of this general topic. The author notes that of all the challenges America's military faces, the one least susceptible to a fiscal or technological fix is geography. For over a hundred years, two great oceans protected the United States from attack. In the world of 2016, they are a major logistical problem for our military to gain access to the battlefield we always want to keep "over there". Transportation solutions are fundamental to that set of challenges. The brilliance of the investment in an Interstate Highway System after WWII has paid both public (including military) and private dividends ever since. Our so-called "lines of communications", including their transportation dimensions, will be subject to interdiction attempts by any adversary. Wartime attrition of those systems becomes a logistician's nightmare. Literature such as *Railway Security* offers valuable insight to strategic planners as well as to thoughtful business leaders. For those who find themselves in significant transportation positions, this book offers the chance to take stock of rail security today and see how important it is to invest the best minds and talents available to secure our railways of tomorrow. Recognizing that *collaboration* is the key to this security is *step* #1. Acting on that recognition is what will be required. This excellent overview provides the framework to get on with that work.

James M. Loy
Admiral, U.S. Coast Guard (ret.)
Former Deputy Secretary, Department of Homeland Security

Preface

The September 11, 2001, attacks were a wake-up call for government officials, operators, and users of all modes of transportation. In the immediate aftermath of the horrifying events of that beautiful late summer day, the nation's railroads, freight and passenger, went immediately on high alert, not knowing if they were to be targeted along with air carriers. The shutdown of the nation's commercial air industry also showed how important it might be to have alternative modes of travel, with intercity passenger rail a clear option. As it became apparent that rail was not a target, attention focused on policies and programs to defend against further attacks on airlines. Rail became somewhat forgotten, with the rail industry largely left to come up with its own approach to securing itself, albeit sharing information on threats with the relevant government agencies. However, attacks on trains and stations in other nations created a public awareness for the need to pay attention to securing both freight and passenger operations.

This book began as a result of the collaboration of authors Plant and Young on several research projects – and many lunchtime conversations – on the need for an interdisciplinary and comprehensive look at the risks and vulnerabilities of the nation's rail system. Jeremy Plant brings a public policy perspective to the project, as well as a long-standing interest in railroads as a subject for photography and historical scholarship and leadership of an academic program in Homeland Security. Richard Young brings both a wealth of practical experience as a shipper using the railroads to move chemicals and an academic perspective on supply-chain management and the interface of rail and maritime modes of transport.

As the project developed, a third perspective, transportation engineering, was added, as Gary Gordon joined the group. He brings not only technical expertise on rail infrastructure but years of experience working on rail security with the Transportation Security Administration (TSA). The net result, we feel, is a book that brings an appropriate interdisciplinary approach to the subject.

We make a number of assumptions in the book, the most important perhaps being the realization that, unless extreme events shift focus from other aspects of homeland security to railroad security, securing the nation's rail network will be a stepchild to the major emphasis on airline security (and perhaps maritime and port security as well). We assume, therefore, that the current emphasis on collaboration among a wide network of government agencies, rail operators, industry representatives, and policing units will continue to be the primary approach used, and embodied in public policy, to secure the rails. Collaboration is based on information sharing and a sense of the proper division of labor and assignment of tasks among the members of the rail security network.

A second assumption is that the approach to rail security will be reactive to changes in the external environment. Change will be incremental but episodic, as major events such as the bombing of commuter trains in Madrid in 2004 and the London transit system in 2005 heighten awareness of risks and create demand for greater attention to common risks and vulnerabilities. But even without dramatic events, we assume a slow but incremental development will persist as organizations involved in rail security continue to learn and evolve, generate more sophisticated models and scenarios, and analyze greater amounts of information on threats and actual attacks.

While our primary focus is on security from terrorism, it is also important to assume that rail security will be influenced by the other sorts of catastrophic events that impact rail operations: non-terrorist human-caused catastrophes, and catastrophic events caused by nature. For this reason, policies and programs for rail safety and emergency response are touched upon as they interconnect with rail security, but our emphasis remains on the possibility that railroads will be the targets of terrorist attacks.

The intended audience of the book is a varied mix of academics from several disciplines and a wide range of practitioners: rail operators, rail shippers, first responders, government officials, and the thousands of travelers who ride the rails. Just as the U.S. rail system is vast and multi-dimensional, so is the topic of rail security. We hope to bring some sense of the importance of the topic as we continue to live in what has been termed "The New Normal": living with the constant and apparently endless threat of terrorism directed at our society.

Abbreviations

3-R Act	Regional Rail Reorganization Act of 1973
4-R Act	Rail Revitalization and Regulatory Act of 1976
AAR	Association of American Railroads
ABS	Automatic Block Signaling
ACC	American Chemical Council
APTA	American Public Transportation Association
ARRA	American Recovery and Reinvestment Act
ASLRRA	American Short Line and Regional Railroad Association
ASSE	American Society of Safety Engineers
BLET	Brotherhood of Locomotive Engineers and Trainmen
BNSF	Burlington Northern Santa Fe
BRF	Bureau of Rail Freight
BZPP	Buffer Zone Protection Program
C&S	Communications and Signaling
CAEPG	Community Awareness Emergency Planning Guide
CBP	U.S. Customs and Border Protection
CBR	Chemical, Biological or Radiological
CCIPP	Commonwealth Critical Infrastructure Protection Program
CERT/CC	Computer Emergency Response Team / Coordination Center
CFATS	Chemical Facility Anti-terrorism Standards
CFR	Code of Federal Regulations
CHEMTREC	Chemical Transportation Emergency Center
CII	Critical Infrastructure Information
CIKR	Critical Infrastructure and Key Resources
CIPAC	Critical Infrastructure Partnership Advisory Council
COFC	Container on Flatcar
COI	Community of Interest
CONRAIL	Consolidated Rail Corporation
COOP	Continuity of Operations
CPC	Casualty Prevention Center
CPTED	Crime Prevention through Environmental Design
CSI	Container Security Initiative
CSIA IWG	Cyber Security and Information Assurance Council Interagency Working Group

CSIRT	Computer Security Incident Response Team
CTC	Computerized Track Control (also Centralized Traffic Control)
C-TPAT	Customs-Trade Partnership Against Terrorism
CWIN	Cyber Warning Information Network
CWR	Continuously Welded Rail
DCS	Digital Control Systems
DHS	Department of Homeland Security
DOC	Department of Commerce
DOD	Department of Defense
DOE	Department of Energy
DOJ	Department of Justice
DOL	Department of Labor
DOS	Department of State
DOT	Department of Transportation
E.O.	Executive Order
ECTF	Electronic Crimes Task Force
EIA	Energy Information Administration
EMPG	Emergency Management Performance Group
EMS	Emergency Management Services
EOC	Emergency Operations Center
EPA	Environmental Protection Agency
ERRI	Emergency Response to Railroad Incidents
EXIS	Exercise Information System
FAF	Freight Analysis Framework
FBI	Federal Bureau of Investigation
FEMA	Federal Emergency Management Agency
FERC	Federal Energy Regulatory Commission
FIRST	Forum of Incident Response and Security
FISMA	Federal Information Security Management Agency
FRA	Federal Railroad Administration
FRSGP	Freight Rail Security Grant Program
FTA	Federal Transit Administration
FY	Fiscal Year
GAO	Government Accountability Office
GFIRST	Government Forum of Incident Response and Security Teams
GPS	Global Positioning System
Hazmat	Hazardous Materials
HHS	Department of Health and Human Services
HITRAC	Homeland Infrastructure Threat and Risk Analysis Center
HMGP	Hazard Mitigation Grant Program
HMR	Hazardous Material Regulation
HSAC	Homeland Security Advisory Council
HSAS	Homeland Security Advisory System
HSC	Homeland Security Council
HSEEP	Homeland Security Exercise and Evaluation Program
HSGP	Homeland Security Grant Program

HSIN	Homeland Security Information Network
HSIN-CS	Homeland Security Information Network for Critical Sectors
HSIP	Homeland Security Infrastructure Program
HSOC	Homeland Security Operations Center
HSPD-7	Homeland Security Presidential Directive 7
HTUA	High Threat Urban Area
IBC	Intermediate Bulk Containers
ICC	Interstate Commerce Commission
IDM	Infrastructure Data Management
IDW	Infrastructure Data Warehouse
IECGP	Interoperable Emergency Communications Grant Program
IED	Improvised Explosive Device
IFR	Interim Final Rule
IICD	Infrastructure Information Collection Division
IICP	Infrastructure Information Collection Program
IICS	Infrastructure Information Collection System
IICV	Infrastructure Information Collection and Visualization
IP	Office of Infrastructure Protection
IRAPP	Infrastructure Risk Analysis Partnership Program
IRC Act	Implementing Recommendations of the 911 Commission
ISAC	Information Sharing and Analysis Center
ISE	Infrastructure Safety and Environment
ISO	International Standards Organization
ISO tank	Independent System Operator Tank
ISTEA	Intermodal Surface Transportation Efficiency Act
I-STEP	Intermodal Security Exercise Program
ITS	Intelligent Transportation System
IV	Infrastructure Visualization
IWWN	International Watch and Warning Network
JTTF	Joint Terrorism Task Force
LEO	Law Enforcement Online
LOS	Level of Service
LPG	Liquefied Petroleum Gas
MARC	Maryland Area Rail Commuter
MBTA	Massachusetts Bay Transit Authority
MIFC	Maritime Intelligence Fusion Center
MIPT	Memorial Institute for the Prevention of Terrorism
MMRS	Metropolitan Medical Response System
MS-ISAC	Multi-state Information Sharing and Analysis Center
N/S	No Signal
NARP	National Association of Railroad Passengers
NARS	North American Rail Shippers Association
NCC	National Coordination Center for Telecommunications
NCIF R&D	National Critical Infrastructure Protection – Research and Development
NCRCG	National Cyber Response Coordination Group

NCS	National Communications Systems
NCSA	National Cyber Security Alliance
NCSD	DHS National Cyber Security Division
NCTC	National Counterterrorism Center
NGN	Next Generation Network
NIAC	National Infrastructure Advisory Council
NIAP	National Infrastructure Assurance Partnership
NICC	National Infrastructure Coordinating Center
NIMS	National Incident Management System
NIPP	National Infrastructure Protection Plan
NISAC	National Infrastructure Simulation and Analysis Center
NIST	National Institute of Standards and Technology
NJT	New Jersey Transit
NJTTF	National Joint Terrorism Task Force
NOC	National Operations Center
NPR	National Performance Review
NRC	Nuclear Regulatory Commission
NRCMA	National Railroad Construction and Maintenance Association
NRCC	National Response Coordination Center
NRF	National Response Framework
NRIC	Network Reliability and Interoperability Council
NRSA	Northeast Rail Service Act of 1981
NSEP	National Security and Emergency Preparedness
NSA	National Security Agency
NSC	National Security Council
NSGP	National Security Grant Program
NSTAC	National Security Telecommunications Advisory Committee
NSTC	National Science and Technology Council
NTD	National Transit Database
NTS	National Transportation Statistics
NTSB	National Transportation Safety Board
NVOCC	Non-vessel Operating Common Carrier
O&M	Operations and Maintenance
ODP	Office of Domestic Preparedness
OHS	Office of Homeland Security
OI&A	Office of Intelligence and Analysis
OMB	Office of Management and Budget
OST	Office of Secretary of Transportation
OSTP	Office of Science and Technology Policy
PATCO	Port Authority Transportation Corporation
PCC	Policy Coordination Committee
PCII	Protected Critical Infrastructure Information
PCIPB	President's Critical Infrastructure Protection Board
PDD	Presidential Decision Directive
PEMA	Pennsylvania Emergency Management Agency

PHMSA	Pipeline and Hazardous Materials Safety Administration
PIH	Poison Inhalation Hazard
PNT	Position, Navigation and Timing
PortSEP	Port Security Exercise Program
PSA	Protective Security Advisor
PSGP	Port Security Grant Program
PTC	Positive Train Control
PVTSAC	Private Sector Senior Advisory Committee
RCRMS	Rail Corridor Risk Management System
RCCC	Regional Consortium Coordinating Council
RCPGP	Regional Catastrophic Preparedness Grant Program
RFAP	Rail Freight Assistance Program
RFID	Radio Frequency Identification
RISS	Regional Information Sharing Systems
RLA	Railway Labor Act
RRIF	Railroad Rehabilitation and Improvement Financing
RSAC	Railroad Safety Advisory Committee
RSI	Railway Supply Institute
RSSM	Rail Security-Sensitive Materials
RTAP	Rail Transportation Assistance Program
RVP	Reid Vapor Pressure Index
SAFETEA	Safe, Accountable, Flexible, Efficient Transportation Equity Act
SAFETY	Support Anti-terrorism by Fostering Effective Technology
SCADA	Supervisory Control and Data Acquisition Systems
SCC	Sector Coordination Councils
SCOR	Supply Chain Operations Reference
SCTG	Standard Classification for Transported Goods
SEPTA	Southeast Pennsylvania Transportation Authority
SHIRA	Strategic Homeland Infrastructure Risk Analysis
SHSP	State Homeland Security Program
SLRR	Short Line Railroad
SPEED	Systematic Project Expediting Environmental Decision-Making
SPP	Sector-Specific Plan
SSI	Security Sensitive Information
SSMP	Safety and Security Management Plan
STAT	Strategic Transportation Asset Tracking
STB	Surface Transportation Board
STRASCNET	Strategic Rail Corridor Network (DOD)
STSIP	Surface Transportation Security Inspection Program
SVA	Security Vulnerability Assessment
TCC	Tank Car Committee (AAR)
TCL	Target Capabilities List
TCP	Transmission Control Protocol
TCS	Traffic Control System

TEA-21	Transportation Equity Act for 21st Century
THSGP	Tribal Homeland Security Grant Program
TIGER	Transportation Investment Generating Economic Recovery
TIH	Toxic Inhalation Hazard
TIR	*Transports Internationaux Routiers*
TOFC	Trailer on Flatcar
T/PIH	Toxic/Poison Inhalation Hazard
TRANSCAER	Transportation Community Awareness and Emergency Response
TRB	Transportation Research Board
TSA	Transportation Security Administration
TSAR	Transportation Statistics Annual Report
TSC	Total Security Culture
TSGCC	Transportation Sector Government Coordinating Council
TSGP	Transportation Security Grant Program
TSI	Transportation Security Incident
TSOC	Transportation Security Operations Center
TST	Transportation, Space, and Technology Center
TVA	Threat and Vulnerability Assessments
TWC	Track Warrant Control
TWIC	Transportation Worker Identification Credential
UASI	Urban Area Security Initiative
UCNI	Unclassified Controlled Nuclear Information
URCS	Uniform Rail Costing System
USACE	U.S. Corps of Army Engineers
USAPA	USA PATRIOT Act
US-CERT	U.S. Computer Emergency Readiness Team
USCG	U.S. Coast Guard
USDA	U.S. Dept. of Agriculture
USDOT	U.S. Dept. of Transportation
VBIED	Vehicle Borne Improvised Explosive Device
VIPR	Visible Intermodal Prevention and Response
VISAT	Vulnerability Identification Self-Assessment Tool
WMD	Weapon of Mass Destruction
WTI	West Texas Intermediate

Glossary

Alighting The act of getting off a train at the passenger's destination.

American Recovery and Reinvestment Act A law enacted as a result of the economic crisis. Its goals are to create new jobs and save existing ones, spur economic activity and invest in long-term growth, and foster unprecedented levels of accountability and transparency in government spending through tax cuts, benefits, grants, and loans.

American Short Line and Regional Railroad Association Trade association of the smaller railroads of which there are more than 550, albeit that Genesee & Wyoming Corporation holds over 100.

AMTRAK Branded name of the National Railroad Passenger Corporation, which was created by the Rail Passenger Service Act of 1970 and given the responsibility for the operation of intercity, as distinct from suburban, passenger trains between points designated by the Secretary of Transportation.

Association of American Railroads The primary trade association for railroads operating in the U.S. – note that Mexican and Canadian railroads with U.S. operations are also members. Also known as AAR.

Auto Train Service East Coast Amtrak service that includes special railcars for transporting passenger-owned automobiles within passenger train consists.

Automatic Block Signaling A signal system that controls when a train can advance into the next track block.

Average Haul The average distance, in miles, one ton is carried. It is computed by dividing ton-miles by tons of freight originated.

Aviation A term encompassing an entire mode that includes aircraft, air traffic control systems, and approximately 450 commercial airports and 19,000 additional airfields.

Block A section of a railroad controlled by a signal.

Blue Flag A marker on a railcar indicating to other personnel that it should not be moved often, because workers are making repairs to track or railcars.

Boarding The act of getting on a train.

Buffer Zone Protection Program Provides funding to increase the preparedness capabilities of jurisdictions responsible for the safety and security of communities surrounding critical infrastructure and key resource assets.

Bulk Train Service For grain, coal, and similar bulk commodities moving in unit trains.

Caboose An obsolete term describing a railcar that was usually placed at the end of a train to accommodate crew members.

Chock An implement that prevents train wheels from turning.

Class I Railroads Line haul freight railroads with 2007 operating revenues of at least $433 million. These are the largest operating in North America.

Classification The arranging of freight cars into trains bound for specific destinations. Action occurs in a classification yard.

COFC Container on flatcar. Usually means that an ocean or domestic container is being transported by rail without wheeled chassis.

Commodity An economic good: as an article of commerce especially when delivered for shipment.

Commonwealth Critical Infrastructure Protection Program A Pennsylvania collaboration between the State Police, the PA Office of Homeland Security, and the Pennsylvania Emergency Management Agency to protect rail transportation.

Commuter Rail Urban passenger train service for short-distance travel between a central city and adjacent suburbs. Does not include rapid rail transit or light rail transit service.

Congestion Cost Value of travel time delay (estimated at $13.45 per hour of person travel and $71.05 per hour of truck travel) and excess fuel consumption (because it is estimated using the average cost per gallon by state, the number will change frequently).

Consist That combination of locomotives and railcars comprising a train.

Container A box-like device used to store, protect, and handle a number of packages or items as a unit of transit that can be interchanged between trucks, trains, and ships without rehandling the contents.

Contracted Service Transportation service provided to a public transit agency or governmental unit from a public or private transportation provider based on a written contract.

CREATE Program A $1.5 billion project to build new overpasses and to modernize signals in the Chicago metropolitan area.

Critical Infrastructure The assets, systems, and networks, whether physical or virtual, so vital to the United States that their incapacitation or destruction would have a debilitating effect on security, national economic security, public health or safety, or any combination thereof.

Cut (of Cars) Two or more freight cars loaded and tendered at the same time and destined for a single consignee. It is of lesser quantity than a unit train.

Dark Territory Segment of a railroad where there is no signaling system.

Diamond An at-grade crossing of two railroad lines.

Double-Stack Cars Intermodal service characterized by shipping containers that are stacked two-high on railcars.

Drill A local train that picks up and delivers freight cars to industries.

Energy Efficiency The ratio of energy inputs to outputs from a process – for example, miles traveled per gallon of fuel (mpg).

Fatality (Rail) (1) Death of any person from an injury within 30 days of the accident or incident (may include non-train accidents or incidents); or (2) death of a railroad employee from an occupational illness within 365 days after the occupational illness was diagnosed by a physician.

Federal Employers' Liability Act A statute that compensates railroad workers injured on the job.

Federal Railroad Administration The purpose of this administration is to promote and enforce rail safety regulations; administer railroad assistance programs; conduct research and development in support of improved railroad safety and national rail transportation policy; provide for the rehabilitation of the Northeast Corridor rail passenger service; and consolidate government support of rail transportation activities.

Federal Transit Authority Allocates funds to support a variety of locally planned, constructed, and operated public transportation systems throughout the United States. Transportation systems typically include buses, subways, light rail, commuter rail, streetcars, monorail, passenger ferry boats, inclined railways, or people movers.

Fossil Fuels Any naturally occurring organic fuel formed in the earth's crust, such as petroleum, coal, and natural gas.

Freight Goods being moved by transportation lines from one place to another.

Freight Cars in Service For railroads, this includes railroad-owned cars plus cars with a railroad's mark, which usually have a longer-term lease. The railroads also use equipment controlled by shippers and leasing companies

Freight Rail Security Grant Program Federal grant to protect critical surface transportation infrastructure from acts of terrorism, major disasters, and other emergencies.

Freight Revenue (Rail) Revenue from the transportation of freight and from the exercise of transit, stop-off, diversion, and reconsignment privileges as provided for in tariffs.

Government Accountability Office A staff unit that supports Congress in meeting its constitutional responsibilities and helps improve the performance and accountability of the federal government for the benefit of the American people.

Government Ownership A railroad that is a government entity or is majority-owned by a government entity.

Grade Crossing An intersection of a roadway and a railroad line. May be either protected or unprotected.

Hazardous Materials Any substance or material that could adversely affect the safety of the public, handlers, or carriers during transportation. Major categories include combustible, corrosive, explosive, flammable, oxidizer, poison, and radioactive.

Hazardous Materials Transportation Act A statute regulating hazardous materials transportation in the United States.

Headway The physical space between two trains.

Heavy Rail A railway with the capacity to transport a heavy volume of passenger or freight traffic and characterized by exclusive rights-of-way, multicar trains, high speed, rapid acceleration, sophisticated signaling, and high-platform loading. It is differentiated from light rail, which designates street cars, subways, and some interurban passenger operations.

High Threat Urban Area Describes cities with both high populations and high population densities.

Highway-Rail Grade-Crossing A location where one or more railroad tracks are crossed by a public highway, road, street, or private roadway at grade, including sidewalks and pathways at (or associated with) the crossing.

Highway-Rail Grade-Crossing Accident/Incident Impact between on-track rail equipment and a highway user (any mode of surface transportation) at a designated crossing site.

Homeland Security Presidential Directive 7 (HSPD-7) U.S. policy for enhancing CIKR protection by establishing a framework for NIPP partners to identify, prioritize, and protect the nation's CIKR from terrorist attacks.

Improvised Explosive Devise (IED) Devices placed or fabricated in an improvised manner incorporating destructive, lethal, noxious, pyrotechnic, or incendiary chemicals, designed to destroy, disfigure, distract, or harass. They may incorporate military stores but are normally devised from non-military components.

Infrastructure (1) An underlying base or foundation especially for an organization or system.

(2) The basic facilities, services, and installations needed for the functioning of a community or society, such as transportation and communications systems, water and power lines, and public institutions, including schools, post offices, and prisons.

Interchange Traffic Receiving rail traffic from other rail companies on company-owned tracks.

Intermodal Transportation activities involving more than one mode of transportation, including transportation connections, choices, cooperation, and coordination of various modes.

Intermodal Train Service For commodities moving in containers or truck trailers on flat cars or specialized intermodal cars.

Interstate Commerce Act of 1887 The act that created the Interstate Commerce Commission. Which was formed to regulate and oversee the railroad system. The commission was abolished in 1995.

ISO Tank Also known as a tank container. They normally come as 20 ft. units that look like a tank inside a cage. They are transported at sea by container ship and by placing them on a chassis that can be moved by truck. When being moved by rail, the container is placed either in the well of a double-stack unit or on a flatcar as COFC.

Key Resources Publicly or privately controlled resources essential to the minimal operations of the economy and government.

Key Train A train consisting of 20 or more cars carrying hazardous materials, five or more cars of TIH, or one car of radioactive material. Key trains have reduced speed limits and need to have their risk assessed.

Lading A load; cargo; freight.

Light Rail A streetcar-type vehicle operated on city streets or underground; it may be on semi-exclusive rights-of-way or exclusive rights-of-way. Service may be provided by step-entry vehicles or by level boarding.

Line Haul Railroads in the business of providing transportation between two points.

Links and Nodes A means whereby logisticians evaluate transportation networks with nodes making reference to fixed points of geography such as stations, terminals, and interchange points. With freight transportation, this can also make reference to both supplier and customer locations. Links, on the other hand, refers to transportation lines that connect nodes.

Locomotive Railroad vehicle equipped with flanged wheels for use on railroad tracks, powered directly by electricity, steam, or fossil fuel, and used to move other railroad rolling equipment.

Logistics The aspect of business operations that deals with the procurement, transportation, warehousing, and distribution of materials and products. The term was originally applied to military operations engaged in similar activities.

Maritime Transportation System For the U.S., consists of about 95,000 miles of coastline, 361 ports, over 10,000 miles of navigable waterways, 3.4 million square miles of Exclusive Economic Zone to secure, and intermodal landside connections, which allow the various modes of transportation to move people and goods to, from, and on the water.

Mass Transit Includes multiple-occupancy vehicles, such as transit buses, trolleybuses, vanpools, ferryboats, monorails, heavy (subway) and light rail, automated guide way transit, inclined planes, and cable cars designed to transport customers on local and regional routes.

Multiple Unit (MU) The connecting together of two or more electric or diesel locomotives so that they can function together under a single control.

National Infrastructure Protection Plan Provides the unifying structure for the integration of a wide range of efforts for the enhanced protection and resiliency of the nation's critical infrastructure and key resources (CIKR) into a single national program. The goal of the NIPP is to build a safer, more secure, and more resilient America by preventing, deterring, neutralizing, or mitigating the effects of deliberate efforts by terrorists to destroy, incapacitate, or exploit elements of our nation's CIKR and to strengthen national preparedness, timely response, and rapid recovery of CIKR in the event of an attack, natural disaster, or other emergency.

National Strategy for Physical Protection of Critical Infrastructure and Key Assets Identifies a clear set of national goals and objectives and outlines the guiding principles that will underpin efforts to secure the infrastructures and assets vital to national security, governance, public health and safety, economy, and public confidence. This strategy also provides a unifying organization and identifies specific initiatives to drive near-term national protection priorities and inform the resource allocation process. It establishes a foundation for building and fostering the cooperative environment in which government, industry, and private citizens can carry out their respective protection responsibilities effectively and efficiently.

National Strategy to Secure Cyberspace It provides direction to the federal government departments and agencies that have roles in cyberspace security.

It also identifies steps that state and local governments, private companies and organizations, and individual Americans can take to improve our collective cyber security.

Non-trespassers (Rail) A person lawfully on any part of railroad property used in railroad operations or a person adjacent to railroad premises when injured as the result of railroad operations.

Non-train Incident Event not involving on-track equipment that results in a reportable casualty.

Northeast Rail Service Act of 1981 Relieved the Conrail and others of their commuter rail operations and fostered the creation of Metro-North, SEPTA, NJTransit Rail, and MARC as regional rail authorities.

Operating Expenses (Rail) Expenses of furnishing transportation services, including maintenance and depreciation of the plant used in the service.

Passing Siding Location where a higher-priority train may pass a lower-priority train.

Pipeline Systems These include vast networks of pipeline that traverse hundreds of thousands of miles throughout the country, carrying nearly all of the nation's natural gas and about 65 percent of hazardous liquids, as well as various chemicals.

Poison Inhalation Hazard Materials which are known or presumed on the basis of tests to be toxic to humans and pose a hazard to health in the event of a release during transportation, such as chlorine and anhydrous ammonia.

Positive Train Control Consist of sensors, location and communications equipment, and advanced computing systems that together provide detailed and accurate information regarding the status of operations on a rail network.

Preliminary National Rail Plan Delivered to Congress on October 16, 2009, the PNRP represents a springboard for development of the long-range NRP by illustrating the role that rail plays in meeting the strategic goals established by the Secretary of Transportation. The preliminary plan lays the groundwork for developing policies to improve the U.S. transportation system. Its goals are consistent with the goals of the U.S. Department of Transportation: to improve safety, to foster livable communities, to increase the economic competitiveness of the United States, and to promote sustainable transportation.

Private or Other Owner Railroad or its equipment that is not owned by a firm engaged in for hire transportation activities.

Private Sector Consists of those entities which are not controlled by the state: i.e. a variety of entities such as private firms and companies, corporations, private banks, or non-governmental organizations

Public-Private Partnership Arrangement where a public agency and a private entity share common interests and agree to cooperate/collaborate for mutual benefit.

Public Sector Part of economic and administrative life that deals with the delivery of goods and services by and for the government, whether national, regional, or local/municipal.

Rail Corridor A restricted tract of land for the passage of trains.

Rail Corridor Risk Management System An algorithm for assessing the relative risk of trains operating over a specific length of track. It was designed and implemented by the FRA to find the route of minimum risk for trains carrying hazardous materials.

Rail Freight Assistance Program Provides financial assistance for investment in rail freight infrastructure.

Rail Freight Infrastructure This includes the rights-of-way, track structure, stations and depots, terminals, intermodal facilities, repair facilities, bridges and culverts, tunnels, communications and signaling, rail yards, control centers, and office buildings.

Rail Passenger Service Act of 1970 Established Amtrak to relieve the major railroads of the financial burden of operating interstate passenger service.

Rail Revitalization and Regulatory Act of 1976 Consolidated the six bankrupt Northeastern railroads into a single railroad called the Consolidated Rail Corporation, or Conrail for short.

Rail Transportation Assistance Program A federal program providing annual grants for financial assistance for investment in rail freight infrastructure. The intent of the program is to (1) preserve essential rail freight service where economically feasible and (2) preserve or stimulate economic development through the generation of new or expanded rail freight service.

Railroad Capacity A term determined by many factors, including the amount of railroad track and rolling stock, the number and power of locomotives, maintenance, staffing levels, and a wide variety of operating strategies.

Railroad Police A law enforcement unit operated by individual railroads but chartered by respective state legislatures to protect passengers, freight, and railroad property.

Railroad Rehabilitation and Improvement Financing Program A federal law establishing low-interest federal loans and credit for railroad infrastructure projects and debt refinancing.

Railroad Retirement Act The federal law that preceded the Social Security Act (49 Stat. 620); it provides a unique system for railroad retirement benefits outside the framework of Social Security.

Railroad Safety Enhancement Act of 2008 A federal law that requires Class I railroads to implement positive train controls on lines carrying hazardous materials and lines carrying both freight and passengers by 2015.

Railroad Unemployment Insurance Act A federal law that provides a special framework for unemployment benefits for railroad workers.

Ramp (aka Intermodal Ramp) A facility for loading and unloading COFC and TOFC railcars.

Rapid Transit Rail Transit service using railcars driven by electricity usually drawn from a third rail, configured for passenger traffic, and usually operated on exclusive rights-of-way. It generally uses longer trains and has longer station spacing than light rail.

Regional Rail Reorganization Act of 1973 Provided financial assistance for continued operations of railroads in the Northeast, many of which were in bankruptcy.

Regional Railroad A railroad that is at least 350 miles and/or has a revenue between $40 million and the Class I threshold ($359.6 million). Also known as Class II railroads.

Right-of-Way The land upon which the railroad tracks are laid.

Rolling Stock Rolling stock comprises all the vehicles that move on a railway. It usually includes both powered and unpowered vehicles – for example, locomotives, railroad cars, coaches, and wagons.

Shipper Ownership Owners ship at least 50 percent of the railroad's carloads.

Shortline Railroads that have less than 350 miles and has a revenue less than $40 million and primarily hauls freight over main or branch line tracks. Also known as Regional Railroads.

Short Ton A unit of weight equal to 2,000 pounds, as opposed to a long ton or metric ton, which is 2,204 pounds.

Slack A measure that is the difference in the length of a railcar when under acceleration versus at rest. Slack action is what allows a heavy train to get moving because it applies force on one car at a time beginning with the first one in the consist.

Staggers Act The deregulation on the railroad industry, providing the industry increased flexibility to adjust their rates and tailor services to meet shipper needs and their own revenue requirements.

State Homeland Security Program Provides funds to build capabilities at the state and local levels and to implement the goals and objectives included in state homeland security strategies and initiatives in their State Preparedness Report. States are required to ensure that at least 25 percent of SHSP-appropriated funds are dedicated towards law-enforcement terrorism-prevention-oriented planning, organization, training, exercise, and equipment activities, including those activities which support the development and operation of fusion centers.

State Railroad Infrastructure Act of 2004 Within Pennsylvania, an act that established a state railroad bank and that allows the Pennsylvania DOT Bureau of Rail Freight, Ports, and Waterways to provide economic development grants through the Rail Freight Assistance Program and the Rail Transportation Assistance Program.

Supply Chain A chain that (1) is a group of buying and selling firms that are dependent upon one another, (2) has the business activities of source, make, and deliver with an overarching or coordinating plan activity, and (3) has three flows transcending all of the members of the supply chain – namely, physical, informational, and financial flows. The term is often ascribed to Robert Stallkamp of Chrysler Corporation in the early 1980s.

Switch A junction of two or more tracks that has a mechanical means to divert trains in multiple directions.

Switching and Terminal Railroad Often engaged in the movement of railcars to specific locations for either loading or unloading, these railroads are generally very short-distance haulers and not considered line haul operators.

Terrorism Acts intending to harm noncombatants, but may include damage to infrastructure critical for economic activity. This is what makes it different from fighting in war. Second, terrorists use violence for dramatic purpose: usually to instill fear in the targeted population. This deliberate evocation of dread is what sets terrorism apart from simple murder or assault.

TOFC Trailer on flatcar. Highway trailers and containers when they are mounted on a wheeled chassis and loaded on railcars.

Ton-Miles The movement of one ton of cargo the distance of one statute mile. For example: two ton-miles could be two tons moved one mile or one ton moved two miles.

Toxic by Inhalation Hazard Risk Reduction Program The program that seeks to establish secure storage areas for TIH materials and to expedite their movement through the system.

Track Block A section of track with traffic control signals at each end.

Track Warrant Control A basic train control system that requires the train crew to obtain written permission known as warrants before entering a section of track.

Trackage Rights The authority of one railroad to operate over the tracks of another railroad for a fee.

Traffic Control System System that use electrical circuits in the tracks to monitor the location of trains, allowing railroad dispatchers to control train movements from a remote location, typically a central dispatching office.

Train Accident Collision, derailment, fire, etc., involving on-track equipment.

Train Incident Event involving on-track equipment that results in a reportable casualty.

Train-Mile The movement of a train, which can consist of many cars, the distance of one mile. A train-mile differs from a vehicle-mile, which is the movement of one car (vehicle) the distance of one mile. A ten-car (vehicle) train traveling one mile is measured as one train-mile and ten vehicle miles.

Transit Security Grant Program Creates a sustainable, risk-based effort to protect critical surface transportation infrastructure and the traveling public from acts of terrorism, major disasters, and other emergencies.

Transportation Security Administration (TSA) Unit of the Federal Department of Homeland Security charged with protecting the nation's transportation systems to ensure freedom of movement for people and commerce. Includes law enforcement programs, security programs, security screening, and grants.

Trespasser (Rail) Any person whose presence on railroad property used in railroad operations is prohibited, forbidden, or unlawful.

TSA Rail Security Rule Imposes several new requirements on facilities that ship or receive high-hazard "rail security-sensitive materials."

Unit Train A train consisting of freight cars originating at the same place and all traveling to the same single destination. Often leased to a single shipper or consignee. Common examples are coal trains heading to power plants, but increasingly also intermodal trains.

Weapons of Mass Destruction Weapons capable of inflicting great numbers of human casualties over a large area.

1 Introduction to railroad security

It's 9 p.m. on a long summer evening at an obscure railroad location in the middle of California's Mojave Desert – ironically named West Siberia in this burning desert that never sees snow or frost. Population: zero – as is the case for its neighboring railroad control points, East Siberia and Ash Hill. As the sky darkens, an eastbound intermodal train drifts downgrade and comes to a stop at the cantilever signal bridge. A van approaches on old Route 66, the two-lane road that parallels the Los Angeles–Chicago mainline of the Burlington Northern Santa Fe (BNSF), one of the busiest stretches of railroad in the U.S. with upwards of 100 trains passing each 24 hours. There are many crews on this stretch of desolate railroad – too many for everyone to know one another well. The crew from the eastbound climb down from the cab and prepare to hand the train over to the crew arriving in the van. There were lots of delays today getting out of the Port of Los Angeles–Long Beach and up Cajon Pass, and then a logjam at the big yard at Barstow, some 75 miles or so to the east, so the crew has run out of time to make it to the crew change point at Needles, another 75 miles to the east. The crewmembers exchange quick greetings, and the new crew is soon off. Just another day on America's busiest and arguably most important rail line.

Three time zones east and six hours later, a neatly dressed young man in his twenties waits to board a commuter train outside Philadelphia. Like many of the other daily commuters, he carries a worn leather briefcase and holds a morning paper in his free hand. The 6:05 to Suburban Station in downtown Philadelphia is on time, and the passengers quickly board as the train scurries through the awakening suburbs to its downtown destination. At the next-to-last suburban stop, the young man picks up his paper and quietly exits the car, leaving his briefcase behind.

Meanwhile, as another sunny late summer day warms the cornfields of southern Minnesota, a lone trespasser wanders down the tracks of the unfenced and unguarded yard of the Dakota, Minnesota and Eastern at Waseca. It's a busy time for agricultural traffic in the Midwest, and the yard is crowded with cars loaded with this year's grain harvest. Climbing furtively atop one of the covered hopper cars, he looks around to see if he is noticed, but sees no one in any direction, and pulls out a can of spray paint.

At about the same time, a TSA surface inspector in Houston is conducting a chain of custody inspection of toxic inhalation hazard material (TIH) located in a Port Terminal Railroad Association (PTRA) rail yard. The inspector notes that there is a discrepancy between the TIH rail cars physically observed in the rail yard and those reported in the yard-list paperwork for TIH rail cars present. Eighteen TIH rail cars were counted, but the yardmaster had 19 TIH rail cars on his yard-list documentation.

Is this a regular day of railroading, or are these settings for catastrophic events? The scenarios noted above are suggestive of the immense task of securing America's railroads against manmade and natural disasters. In each of the four cases, things may be simply normal – the crewmembers on the BNSF train in the desert are in fact the rightful employees. The young man has simply forgotten his briefcase. The chemical cars have not been tampered with during their stay at the plant. The trespasser is there to spray paint his signature on the grain car, not to contaminate its contents. And the TSA inspector found that, according to the yardmaster, the "missing" TIH rail car had just departed the rail yard and the paperwork has not "caught up". But all these situations also have the potential to be catastrophic events with immediate and long-term consequences. Imagine the opposite results. The crewmembers boarding the train are in fact part of a domestic terrorist group intent on disrupting American life. They have trained to know the protocols of operating the train and, once started, plant a bomb and jump off the moving train as it glides down the grade to the little crossroads community of Amboy. By the time the bomb explodes, the train is moving at speeds upward of 100 miles per hour and devastates the community, killing a carload of Marines returning from a weekend in Las Vegas to the nearby base at 29 Palms.

The erstwhile commuter is in fact a newly recruited member of an international terrorist organization. Not a suicide bomber – he can live to fight another day, as the commuter train is an easy target, with no security lines to board and minimal attention to someone leaving his briefcase behind. A briefcase loaded either with explosives or anthrax powder, in either event a disaster of great proportions, awaits as the train continues its otherwise unremarkable inbound trip to the metropolis.

The trespasser is no typical graffiti artist but in fact a disgruntled former employee of the big grain company whose privately-owned hopper car he has boarded. The spray paint can is a neat way of concealing his real motive – to contaminate the load of grain with a highly toxic poison. From his former work at the grain elevator, he is familiar with rail cars and how to gain access to the contents inside. His work done, he watches as the train is assembled and moves east along with 100 similar loaded hoppers.

What if the yardmaster could not confirm that the TIH rail car had departed the yard? Where was the TIH rail car? Had it been delivered? Had its markings been changed? Had it departed undocumented? If any of the latter, what was the intent of "hiding" a TIH rail car in the country's fourth-largest and densely populated city?

A third way of looking at the examples is to focus on what may have been done to avert the four catastrophes. In the first case, a background search of BNSF crew members discloses a possible link between a militia group and a driver of one of the vans used to transport crews to various re-crewing destinations. Information shared between the FBI, DHS, and the BNSF police indicate an increased level of communication between the suspect and a group on the terrorist watch list. The railroad police notify the California Highway Patrol, and both are in place and apprehend the bogus crew as they prepare to board the train.

As the commuter train leaves the station, one of the passengers notices that the young man in the briefcase scenario is acting suspiciously, looking nervously around and repeatedly opening and closing the case and moving it quickly away from another passenger standing close-by on the platform. Without drawing the attention of the man, he notifies the local police from his cell phone, who in turn notifies the head of security for Philadelphia's 30th Street Station, activating a well-rehearsed response by a network of police and first responders. As the man leaves the train, he is quickly brought under custody as a team hurries into the car to retrieve the case and deliver it to the waiting bomb squad.

In the Houston rail yard and through interviewing yard workers, it was learned that the missing TIH rail car had just departed on a train. The train designation and route was known, but the TIH rail car in question was not on the train's manifest. The TSA inspector immediately notified the railroad police, who, in turn, alerted the local police along the route, as well as appropriate government agencies at all levels. Fortunately, the train was stopped at an interlocking not too far from the rail yard and the TIH-carrying rail car was secured. After a brief investigation, it was learned that the conductor on the train was sympathetic to a local gang with Central American ties and intended to release the TIH in an adjacent neighborhood.

The third case, however, is more troublesome. Out in America's heartland, things are open and accessible, and little thought has been given to rail cargo as a way of settling a personal grudge against a former employer. Unless the individual has acted out his anger beforehand, the contaminated cargo is likely to remain undetected, and its effects difficult to trace back to the offender. And in the fourth case, DHS and the FBI have learned that gangs have been recruited by terrorist groups to assist in transporting weapons and people and to support their ideology – for a price.

The purpose of these examples is to point out the immense task facing the network of organizations and individuals charged with securing the nation's rail system against catastrophic events. The U.S. rail system is huge, complicated, and intertwined with other modes of transportation and critical industries. Its infrastructure includes not only track and rolling stock but bridges, tunnels, overpasses, stations, signals and communications systems, and maintenance of way facilities. Its tracks extend through major urban and populated areas, and into chemical plants, grain elevators, port facilities, and factories; its rights-of-way often parallel mass transit systems and heavily traveled highways; share stations with buses, trolleys, and subway trains; and run near places of iconic value, water supplies and large public sporting venues.

The potential for disaster is also clearly real. Looking at terrorism globally, one is struck by how often a train is the target. Mumbai, Moscow, Madrid, London – time and again terrorists have seen passenger rail as an attractive and relatively vulnerable target. The goal may be killing innocent passengers, disrupting the economic and social life of major cities, and crippling the efficient operations of the passenger and freight rail systems – pointing out the inability of governments to protect their citizens. Freight rail does not appear to be as much of a target as passenger rail. But, and as a former TSA intelligence officer stated, the threat to freight rail and, in particular, hazmat trains is "low threat but not a no threat."

It is useful at this point to define *terrorism* and *terrorist*. Jessica Stern (1999, p.11) defined terrorism in this manner:

> Two characteristics are critical for distinguishing terrorism from other forms of violence. First, terrorism is aimed at noncombatants. This is what makes it different from fighting in war. Second, terrorists use violence for dramatic purpose: usually to instill fear in the targeted population. This deliberate evocation of dread is what sets terrorism apart from simple murder or assault.

What Stern's definition clarifies is the basic lesson of the *New Normalcy*: that it is the *acts* and the *motivations* of groups, not the nature of the groups involved, that is most important to understand. Terrorists can be domestic or foreign, working singly or as part of an organized conspiracy, motivated by idealism or hatred, rational or irrational in their goals.

Why do terrorists attack rail targets? Clearly it fits the description of terrorism provided by Stern: it is a way of attacking civilians, it adversely affects the economy, and it is guaranteed to instill fear and demonstrate dramatically the risks involved in riding passenger trains, or being in a crowded station, or even being in a community suddenly contaminated by a hazardous waste spill or chemical release from a derailed or sabotaged freight train.

As tragic and frustrating as attacks on passenger rail have and presumably will continue to be, attacking the freight operations of the sprawling U.S. rail system may be even more chilling in its potential for catastrophe. Freight trains in the U.S. are the preferred mode for the movement of hazardous materials, to include poison and toxic inhalation hazard (P/TIH) material. In recent years the movement of highly volatile crude oil from the Bakken oilfields in the northern plains has increased the threat from catastrophic events involving tank cars. Railroads move increasing amounts of intermodal freight into the nation's ports and bound for destinations inside the country. Rail operations may be the intended target or simply the means of moving a dirty bomb or a vial of anthrax from one point to another. A serious attack on rail freight, especially if it is perpetrated on an intermodal train, has the potential to wreak enormous economic damage, potentially closing off port traffic for a period of time with economic consequences in the billions of lost dollars for the U.S. economy.

Adding to the potential for disaster is the lack of examples to provide an empirical basis for programs to secure the nation's rail system. As a result, policymakers and first responders are left to deal with the most-wicked category of potential risk: low-probability, high-impact events. An alternative is to study events in other national settings, but these run afoul of the consideration that the systems are often fundamentally different as are the organizations and individuals committing acts of violence. Modeling catastrophic events and planning for response and recovery are vital but must remain somewhat removed from a real sense of the probability of rail as a target for terrorist activity. Much can be learned from examining the impact on the rail system of non-terrorist catastrophic events, caused by human error or negligence or by natural disasters, but these inform us more about response and recovery than the critical variable, prevention.

The common element in all these projections is *uncertainty* – about targets, about possible perpetrators, about timing, about single or coordinated attacks. With high levels of uncertainty comes the need for careful analysis of risk. A recent report of the National Research Council's Committee to Review the Department of Homeland Security's Approach to Risk Analysis notes that "although risk analysis is just one input to decision making, it is an essential one" (National Research Council 2010, p. 1), adding that "proper recognition and characterization of both variability and uncertainty are important in all elements of a risk analysis, including effective interpretation of data as they are collected over time on threats, vulnerabilities, consequences, intelligence, and event occurrence" (p. 3).

Risk assessment is still a field of analysis seeking answers to its most pressing questions: how to inform decision makers when empirical data is limited or not directly related to the situation under review; how to coordinate both the analysis and the interpretation of risk assessments among a variety of organizations, public and private, national and local, involved in potentially threatened operations; and how to inform a variety of decisions and decision makers, from high-level policy makers and strategists to those with boots on the ground as first responders (National Research Council 2010).

Approaches to rail security

A little-known event during World War II points out how much has changed in regard to threats to the rail system in the past half-century. On a June evening in 1943, a squad of four German-Americans were landed by a German U-boat at Amagansett, Long Island, with orders to sabotage major elements of the U.S. rail infrastructure. Their targets included the huge Limeville Bridge over the Ohio River on the Chesapeake & Ohio, a critical point on a major rail route for Appalachian coal to fuel the factories of the Midwest, and possibly the famous Horseshoe Curve on the Pennsylvania Railroad, at that time one of the busiest if not the most busy spot on the American rail map. An observant coast guardsman, John C. Cullen, apprehended the group as they came ashore and, instead of being killed by the Germans, was asked to take a bribe and not report the incident. Left alive, Cullen reported the incident, by which time the group had traveled to Manhattan, where they and their leader (Georg Dash, a former waiter) were enjoying themselves, spending money on hotels and liquor rather than their appointed task of sabotage. When apprehended they squealed to the FBI on the second part of the gang, which had landed in Florida, and all eight saboteurs were arrested, tried by a military court as enemy spies, and sentenced to death. Codenamed Operation Pastorius, what started as a serious plan to disrupt the vital U.S. war effort turned first into a comic opera, then tragedy as six of the eight were executed while two others, including Dash, received severe prison terms in exchange for cooperating with U.S. authorities (Persico 2001, pp. 199–205; Plant and Young 2007, p. 10).

Think of how things have changed since 1943:

- The Germans had great difficulty in reaching the U.S., had limited capability to communicate with one another, and relied on conventional means – bombing of critical infrastructure – to disrupt freight operations. Today, terrorists have cell phones, the internet, and other devices that facilitate communications and have the potential to use weapons of mass destruction, not simply conventional guns and bombs.
- Warfare today is not necessarily or even commonly between nations that have declared war on each other. The Germans were clearly identified as enemy agents and subject to the traditional rules of war, leading to their swift arrest, trial, and execution. In the New Normalcy, state sponsorship is not needed or often desired by terrorists belonging to private, non-state terror organizations. Ironically, the level of training, commitment, and expertise employed by such non-state supported terrorists often exceeds that of state agents like the Pastorius gang.
- Critical infrastructure, including major points on the rail system targeted by Operation Pastorius, were guarded 24/7 by uniformed personnel from the U.S. military and National Guard. In the post-9/11 New Normalcy, such intense surveillance is feasible from neither economic nor technological standpoints.
- Although it is hard to attach the label "risk analysis" to wartime efforts to thwart enemy sabotage, the nation's rail system, which during the war moved over 90 percent of the nation's freight and intercity passenger traffic, was clearly seen as among the most critical elements of the nation's infrastructure and given significant resources to defend. Trespassing, photography, or other suspicious action was likely to trigger quick response from civilian and military authorities, railroad personnel, and average citizens. Today, it is unclear how the rail system compares with other inviting targets for terrorist attack or how vigilant our society is regarding threats to the rails.
- The rail system itself has changed dramatically. There are fewer freight railroads, fewer passenger operations, faster trains moving over a streamlined rail system whose mainlines now carry close to capacity numbers of train movements daily. The system, which was actually prohibited from intermodal freight movements by the Interstate Commerce Commission during most of the regulatory era, now functions as a partner with trucking firms, drayage firms at ports, marine operators, and even air operators in an increasingly globalized world of commerce. Securing the rail system means focusing as much on the interchange points and chain of custody as on securing critical rail infrastructure.
- Technology has changed the way railroads operate. Fewer employees – and as a consequence fewer eyes on the ground – are needed to move trains from one point to another. Where in the past, rail personnel were distributed across the rail network in towers, stations, in maintenance of way (MOW) crews, and so on, today's technology allows for signals to be controlled centrally, often time zones away from where the train is actually moving; maintenance

requires fewer workers and employs sophisticated new technology and equipment; and computer software and systems allow many aspects of rail operations to be contracted out to third parties. All of these trends elevate the risk associated with a new type of critical rail infrastructure: the computerized systems that are essential to the safe and secure movement of trains.
- The rail system itself is no longer owned and operated by private companies engaged in both freight and passenger services. Although they may share one another's tracks and other facilities, responsibility and ownership of freight and passenger operations is divided, with private corporations owning the nation's freight railroads and quasi or totally governmental entities such as Amtrak and regional transportation authorities responsible for moving passengers. The freight system is also divided between the seven major trunk lines and over 550 shortline and regional operators, many of which have come into existence as a result of the sale or abandonment of lines formerly owned and operated by major railroads.

The "New Normalcy" means the threats to the rail network cannot be handled in ways that worked in the past. Even in the dark days of 1943, everyone knew that the war would end in a matter of years; that an occasional German or Japanese agent might be landed in the territorial U.S. but that such threats were not existential, in the way that long-range bombers or enemy invasion might be; that the results of any attack would not be targeted to the random murder of civilians or be catastrophic in its impact. All such assumptions now are irrelevant; the world has become a more interconnected and dangerous potential theater for terrorist acts.

Without a triggering event like the 9/11 attacks, there was little motivation to rethink rail security in the years between World War II and the downing of the Twin Towers. That set of assumptions was changed forever as the 9/11 attacks showed the vulnerability of critical infrastructure and transportation systems, and the lack of attention to homeland security issues in the U.S.

Risks and threats to the rail system

The concept of *risk analysis* is at the heart of any approach to secure the homeland. Risk analysis is a method of relating three factors: *vulnerability*, *threats*, and *consequences*. This is the approach taken by the Department of Homeland Security and its network of partners in state and local government and private sector organizations. Vulnerability is based on an objective assessment of the likelihood that terrorist attacks, major disasters, or other emergencies could occur and produce catastrophic consequences. It doesn't take long to think of ways in which rail systems are vulnerable:

- The rail system in the U.S. is enormous, extremely linear, highly complex, and largely unfenced, unguarded, and open to trespass.
- The rail system has a great deal of critical infrastructure, notably tunnels, bridges, causeways, overpasses, and highway crossings – the most frequent estimates say more than 100,000 – and signaling systems.

- The freight rail system is the preferred means of transport for dangerous cargoes, conveyed in specialized rail cars that are easy to identify. Many are labeled as carrying toxic and poisonous materials for reasons of safety in case of damage to aid rail workers and first responders.
- Both freight and passenger rail operations rely heavily on sophisticated signaling, communications, and information systems that may be sabotaged.
- Rail operations connect to other modes of transportation and other commercial and governmental facilities whose security is codependent with that of the rail industry. In particular, ports carrying intermodal cargo, chemical plants, and mass transit systems that share rights-of-way and stations are critically linked to the railroads' vulnerability.
- Although less labor intensive than in the past, railroads employ a large number of individuals and need to insure that there are programs in place to screen employees to insure secure operations.
- Passenger rail operations, especially in and around large metropolitan areas, have a large number of points of entry and exit and rely on quick loading and unloading of passengers to be efficient. This makes the airport model of screening of baggage and passengers infeasible to transfer to rail passenger operations.

Before considering the nature of threats to the rail system, let's quickly identify some likely consequences of major attacks on the railroads, grouping them into those associated most with passenger rail and then freight rail. In considering consequences, it is important to keep in mind that the list must include not simply the consequences intended by the perpetrators of violence but those that stem from the nature of the damage wrought.

Passenger rail consequences are easier to identify than those for freight rail, because of the frequency with which passenger rail has been attacked in recent years in other national settings. These include:

- Loss of life and serious injuries to passengers. Crowded passenger trains typically carry upwards of 1,000 passengers. Loss of life and casualties in bombings, active shooter attacks, derailments, or other sorts of attacks usually will mount into double and often triple figures; injuries will typically be a multiplier of two to four for the number of fatalities. Unlike attacks on commercial aviation, where usually all people aboard are killed, rail attacks provide gruesome and troubling images of dead and wounded, adding to the shock value of such attacks.
- Infecting passengers with communicable diseases. Usually we think of attacks as intended to produce immediate and devastating consequences. However, the unguarded nature of most passenger rail makes it possible to use the crowded train not as a scene of violent and gory death but instead as a way of transferring lethal biological weapons – one thinks immediately of anthrax powder – and using the passengers on the otherwise unremarkable commuter train as agents of death as they move from the station to the office or theater or mall or sports arena or the reverse.

- Shutting down major cities. Whether it is a goal of terrorists or a byproduct of their actions, any attack on a passenger train is likely to have considerable economic consequences, in direct proportion to the role that rail plays in moving people in and out of a given city or community. Given that commuter rail operations are vital to many of the largest cities in the nation – New York, Chicago, Philadelphia, San Francisco, Boston, and Washington as examples – crippling the passenger rail system in such settings has short and long-term consequences for the economic life of the city, region, and nation.
- Affecting modal choice. The problems associated with automotive travel continue to increase – congestion, air quality and other environmental considerations, energy independence – hence, rail has increasingly become the major modal alternative to the automobile for commuting and short intercity trips. Thus, any loss of confidence in the safety and security of rail travel has the potential for serious consequences for individual decisions and public policy direction.

Freight rail attacks, or disruptions caused by non-terrorist human error or natural calamities, create a different but equally compelling list of probable consequences:

- Mass devastation and widespread loss of life. Two likely scenarios for this consequence are attacks on hazardous cargoes carried in freight trains, such as lethal chemicals, and weapons of mass destruction conveyed from overseas in containers and off-loaded onto intermodal trains that move the dirty bomb or biological agent into the nation.
- Serious economic damage. Think again of the dirty bomb in the container scenario, and imagine if the explosion occurs not in some Midwestern city where the train has brought it but in the Port of Los Angeles–Long Beach where it has just been off-loaded. The likely consequence would be shutting down all port operations for a period of time, which in turn would seriously disrupt foreign and domestic commerce and cause major shippers and users untold economic hardship. Walmart and other retailers would be crippled, and the logistics activities of much of the U.S. and world economies would take months, if not years, to recover.
- Tainted cargo. Just as the example above of the passenger train spreading anthrax from infected passengers shows the potential of trains as agents of biological weapons, so does an example, fortunately hypothetical, of a freight train carrying contaminated food or agricultural products. While the bulk of food products move by water or road in this country, a growing percentage is once again handled by freight rail operators, who also carry most grain, animal feed, and other cargoes that if contaminated could enter the food chain with catastrophic effects.
- Environmental damage. Freight trains every day convey dangerous substances across major waterways, through pristine forests and the fragile eco-systems of plains, wetlands, and deserts. Again, even if the intent of the perpetrator has nothing to do with the damage his or her act may cause to the

environment, the consequence of an attack on a bridge or tunnel causing a derailment may be devastating to the environment. In this case, we need only to look to accidents on the rails to show how consequential to the environment such events can be.

As we discuss in the following chapters, the rail system in the U.S. is best thought of as containing three major elements: passenger operation and operators (including Amtrak, which runs the nation's intercity rail system) and regional authorities operating commuter services; the seven major freight carriers, which operate across state boundaries and are best viewed as part of not just a U.S. system but a North American system; and the over 550 regional and short-line railroads that serve as feeders to the big freight railroads and service local businesses and industries (and are sometimes owned and/or operated by the major companies they serve). The consequences of catastrophic events to these different sets of operators vary according to the nature of ownership. The freight system, both trunk lines and shortlines, are privately owned and operated and keenly aware of the advantages they gained through deregulation after 1980. For the first time in living memory, not only are all the major freight operators profitable, but their business is expanding. They fear that a likely consequence of any catastrophic event would be severe economic hardship. Insurance premiums, for instance, might go off the chart. The public might clamor for more regulation. There might be even more requirements of the sort already mandated for changes in the way that the companies conduct business, hiring more specialists in risk assessment or policing. Or hazardous materials might be transferred to modal competitors, trucking firms, and inland waterways, with poorer records of safety, significantly more exposure, and greater risk to the environment.

Passenger operators would be affected differently, but with deleterious consequences of their own right. Amtrak might see its growth in ridership quickly abate and its finances take a dip, allowing its opponents in Congress and elsewhere to call again for its demise. Commuter operators might see both a decline in fare box receipts and an increase in costs associated with the protection of passengers and the prevention of future events, leading in turn to problems of balancing the books without fare increases or tax-funded subsidies.

This leads us next to consider threats to the rail system. To be called a threat, there needs to be two factors in place: *intent* and *capability*. Intent can be determined empirically or through consideration of likely scenarios like the 15 National Planning Scenarios that form the basis of much of the homeland security risk analysis. In the case of natural disasters, the origin is easy to show: hurricanes, floods, and other natural catastrophes can be studied over time and through scientific analysis. In the case of human-caused catastrophes, the intent has to be either shown through empirical data or based on logical deductions in hypothetical situations.

Capability is a measure of the ability of individuals and groups to carry out their intended actions. The next section of the chapter discusses the new homeland security policies and administrative approaches following the 9/11 attacks, and their relevance for rail security.

Rail security and homeland security policy

A number of published works and government reports have reviewed the development of homeland security policy and administration, focusing largely on the national response to the 9/11 catastrophe in passing the USA PATRIOT Act, the Homeland Security Act and the creation of the U.S. Department of Homeland Security, executive orders of the President to direct the new department's activities in particular directions, and subsequent legislation to build on the framework and foundation set in the years immediately following the airline attacks of 9/11. What has been less often reported is the response of state and local governments and private sector organizations to the new reality. The privately owned freight railroads were quick to act, with the industry's two related trade associations – the Association of American Railroads (AAR) and the American Short Line and Regional Railroad Association (ASLRRA) – serving as the central hub for a network approach to sharing information. Five teams of experts were assembled and worked to establish a coordinated approach to deal with four principal concerns: hazardous materials shipments, operational security, military interactions, and technology. The network of involved organizations included not only the private rail operators but also the U.S. Department of Transportation (specifically its units, the Federal Railroad Administration, the Pipeline and Hazardous Materials Safety Administration, and the Federal Transit Administration), Amtrak, and the FBI. After its formation, the Department of Homeland Security was added as a major partner in the network (Plant 2004). For the first few years after the 9/11 attacks, this network of public and private organizations served as the primary operation securing the nation's rail system from threats.

The creation of the Department of Homeland Security through the Homeland Security Act of 2002 set in motion a number of changes, although for several years the situation was in flux as the new department moved to organize its operations and take action on policy initiatives mandated by Congress and the President. Two agencies of DHS were most significant in dealing with railroad security. The Office of Domestic Preparedness (ODP) undertakes risk assessments and provides grants to urban areas under the Urban Area Security Initiative (UASI) program. ODP provided risk assessments for passenger rail and mass transit systems and initially dispensed over $100 million to urban rail systems and $7.1 million for Amtrak through 2005 (Plant and Young 2007, pp. 44–45).

The agency charged to oversee the security of all modes of transportation is the Transportation Security Administration (TSA). TSA was transferred from the Department of Transportation (DOT) to DHS as part of the reorganizations that formed the new department. TSA is empowered by law to deal with the security of all modes of transportation, but its mandate outside commercial air security was largely left unspecified for the first years after its transfer. As noted by the Government Accountability Office (GAO) in 2006,

> ATSA (the Aviation and Transportation Security Act of 2002) does not specify TSA's roles and responsibilities in securing the maritime and land

> transportation modes at the level of detail it does for aviation security. Instead, the act broadly identifies that TSA is responsible for ensuring the security of all modes of transportation.
>
> (GAO 2006, p. 10)

As TSA's surface program evolved, several programs and regulations formed the foundation for rail and surface transportation security. In 2005 in response to the 2004 Madrid train bombings, TSA stood up its Surface Transportation Security Inspection Program (STSIP) with two security directives focusing on passenger rail operations and ATSA as its basis. Without rail-focused regulations, partnered initiatives like the Baseline Assessment for Security Enhancements (BASE) for transit and passenger rail and transit operations, Security Action Items (SAI) focusing on the handling of toxic inhalation hazard (TIH) material rail cars within UASI defined high threat urban areas (HTUA) and Freight Rail Corridor Assessments (FRCA) focusing on the movement of TIH train through the UASI HTUAs.

Evolved from the SAIs and FRCAs, 49 CFR 1580, Rail Transportation Security, was enacted in 2008 and addressed chain of custody of TIH trains, establishment of rail security coordinators, and reporting thresholds and procedures. TSA worked closely with the Federal Railroad Administration (FRA) on this regulation and others established by FRA and enacted a memorandum of understanding that addressed reporting of incidents and other hazardous conditions observed by one party of the other's regulations. An example would be a TSA surface inspector finding a dangerous track condition of immediate concern under FRA regulations (49 CFR 213). There is pending legislation regarding training for front-line operators and vulnerability assessment and security plans for railroads. These and other programs and regulations will be discussed in more detail, as pertinent, in subsequent chapters.

Operationally, the Visible Intermodal Prevention and Response (VIPR) program was established for rail passenger and transit operations. The VIPR program led by TSA's Office of Law Enforcement / Federal Air Marshall Service (OLE/FAMS) is a teaming of law enforcement and other disciplines, such as TSA behavior-detection officers and surface inspectors, to augment transportation security at stations, ports, etc. with the intent of deterring terror-attack planning and execution by creating a randomness of presence and force structure and size at any transportation location at any time. As its success was recognized, it was expanded to include other surface and maritime transportation locations and operations, such as cruise and ferry terminals, bus stations, truck and intermodal terminals, and even airports.

Homeland security policy emanates not just from legislation but also by executive orders of the President. DHS was designated the lead agency for transportation infrastructure protection in Homeland Security Presidential Directive 7 (HSPD-7) issued by the White House on December 17, 2003. This authority was delegated within DHS to TSA. HSPD-7 established national policy for federal departments "to identify and prioritize United States critical infrastructure and key resources and to protect them from terrorist attacks." This directive established the secretary of the Department of Homeland Security "to be responsible for coordinating the overall national effort to enhance the protection of the critical infrastructure and key resources of the United States."

After Michael Chertoff assumed leadership at DHS, he stressed two principles to guide the work of the organization: the use of risk assessment to identify the likely targets of terrorism and their impact, and the need to work within networks of concerned organizations, in both the public and the private sectors. After the London bombings in July 2005, he reiterated his commitment to the network approach:

> We must draw on the strength of our considerable network of assets, functioning as seamlessly as possible with state and local leadership, law enforcement, emergency management personnel, firefighters, the private sector, our international partners, and most certainly, the general public. Building effective partnerships must be core to every mission of DHS.

The sense of partnership to secure critical infrastructure and key resources (CIKR) remains the basis of national homeland security policy. Railroad security falls into the category of transportation systems sector in the DHS list of critical infrastructure, and it is considered one of the most likely targets of terrorist attack, based on the risk analysis that judges vulnerability, threats, consequences, and ultimately risk. Rail security policy continues to evolve as these factors change over time. At the heart of this effort is the effort to determine threats and use such information to inform decisions. In Chapter 3, we look in much greater detail at the development of rail security policy and its implementation.

Seven themes

This volume is intended to give a balanced view of the risks faced by the U.S. rail system in the period of time in which we live – the post-9/11 New Normalcy. It is written with the knowledge that, at any time, on any day, there is the potential for an event that will transform the system in ways we can treat only hypothetically. That said, it is good to start with some shared assumptions:

(1) Rail security efforts will require collaboration and strategic partnerships between federal agencies; state, tribal, and local governments; shippers; ports; truckers; and railway operators. The key factor in successful collaboration will be building trust among the participants and creating and maintaining effective and timely communications among the members of the rail security network.
(2) Rail security efforts cannot be based solely on reactions to catastrophic events, either in the U.S. or abroad. A proactive, planning-based approach is critical to prevent catastrophic events and, if they occur, to react quickly and effectively to mitigate their effects.

(3) Rail security must be seen in the broad context of supply chain management, global trade, and the interconnections between the rail system and other transportation modes and industries.
(4) Catastrophic events fall into three categories: natural, non-terrorist human-caused, and terrorist attacks. While much of the focus of this book is on the potential for terrorist attacks on the nation's rail network, other catastrophes caused by human error or natural disasters or the exploitation thereof provide opportunities to learn about the vulnerabilities of the rail system and effective ways to react and move forward. To this end, issues of rail safety and rail security are inherently linked, even if responsibility for programs and responses are entrusted to different government agencies.
(5) Rail policing must be, and is, moving from traditional approaches to property management and surveillance to information sharing and proactive approaches to working with other members of the rail security network and public agencies.
(6) Although freight and passenger operations are run by separate entities and present different risks and vulnerabilities, the interconnected nature of the rail system requires that a holistic approach be taken and communications between freight and passenger operators assigned a high priority in planning and implementing programs for rail security.
(7) Securing the rail system requires that the system be seen as more than just trains and tracks, but also the infrastructure required for the system to function and the personnel employed to make it work. And don't forget the areas through which they pass. Stations, signals, roadbed, information systems, tunnels, and bridges – infrastructure protection – all fall under the topic of railway security, as does concern the for human resource issues of identity security, screening of employees, and training: to prevent insider threats.

Organization by section and chapter

The book consists of four principal topics spread over 11 chapters. It begins by describing the nature of rail operations as well as some background on how the U.S. rail network has been protected. Clearly, recent times have changed the entire landscape – hence there follows a comprehensive discussion of the threats to the infrastructure.

This chapter has identified the major topics of the book. Chapter 2 describes the rail system in the U.S. and provides a foundation for understanding rail operations and connections to other modes and industries. Chapter 3 examines the nature of railway infrastructure and supporting signal and train control systems. It looks at the issues that infrastructure and signals present for security and how they can be protected beforehand to minimize vulnerability and risk and system redundancy so as to be able to resume operations as soon as feasibly possible after an event. Chapter 4 examines railroad policing and its evolution after 9/11/2001 to its critical role in rail security.

Chapter 5 looks at rail freight operations, including the connections of freight rail with other modes of transport. Chapters 6 and 7 continue with the focus on freight security, by examining the two sorts of freight movements that present the highest risks: hazardous materials and petroleum cargoes, respectively.

Chapters 8 and 9 shift the subject from freight to passenger operations. Chapter 8 examines the burgeoning commuter rail system in the U.S., while Chapter 9 shifts from commuter to intercity passenger operations. As we note, the two types of passenger operations are often in close proximity to one another or even share the same tracks, but significant differences exist between them in risks and vulnerabilities.

Chapter 10 follows with an examination of emergency management and its relationship to security. Preparedness and response is important and differs between natural disasters (which can be forecasted to some extent) and human-caused disasters that have little or no warning. An all-hazards approach is one that will be discussed to address this. Further, security postures must be maintained during natural disasters, as they could be exploited.

Chapter 11 concludes the examination by returning to the major themes of the study, considers where we are currently in securing the railway system, and suggests what the future may hold for rail security. It concludes by looking into the future to identify issues that need to be addressed to insure the security of rail operations. We review the seven themes that run throughout the book (the critical nature of railroads to the functioning of the U.S. economy and society; the connectivity of rail to other major industries and modes of transport; the need to see rail security as something other than rail safety; the differences in security needs and organizational arrangements of the freight and passenger components of the system; and the difficulty of assessing threats to the rail system) to give a sense of what is required to continue and enhance the approach that so far has helped the nation's rail system remain safe, secure, and economically strong.

Bibliography

$1 billion in questions. (2005, July 30). *The Washington Post*, p. A18.

American Short Line and Regional Rail Association. (2006, February 3). *The American short line and regional railroad industry*. Presented to Transportation Security Administration, (mimeo).

Amtrak. (2006). Amtrak operating report. Retrieved January 6, 2007 from www.amtrak.com/servlet/contentserver/pagename=amtrak/am2copy/title_merge_copy_page&=copy&cid=108144267447&ssid=322.htp/october 2006.pdf

Association of America Railroads. (2002, January). *Railroads: The safe way to move*. Washington, DC: Association of America Railroads.

Bogdanich, W. (2004a, July 11). In deaths at rail crossings, missing evidence and silence. *The New York Times*, p. 1.1.

Bogdanich, W. (2004b, July 12). A crossing crash unreported and a family broken by grief. *The New York Times*, p. A1.

Bogdanich, W. (2004c, November 7). For railroads and safety overseer, close ties. *The New York Times*, p. 1.1.

Bogdanich, W., and Drew, C. (2005, January 9). Deadly leak underscores concerns about rail safety. *The New York Times*, p. 1.1.

Breskin, I. (1995, June 21). Lessors expand tankcar fleets. *Chemical Week*, 156(24), 44.

Bureau of Transportation Statistics. (2004). *America's freight transportation gateways*. Washington, DC: U.S. Department of Transportation.

Canadian Transportation Agency. (2002, March 7). *Review of rail security measures post November 11th, 2001*. Paper presented at the Canadian Transportation Agency Meeting on Railway Issues, Ottawa.

Cantrell, B., Lazo, M., and Ruttenberg, R. (2006, November). *Training in hazmat and rail security: Current status and future needs of rail workers and community members*. Retrieved December 11, 2006 via Citizens for Rail Safety Inc. via from www.citizensforrailsafety.org/docs/CRS_PAPER1.doc

Carafano, J. J. (2005, March 18). *The FY 2006 budget request for homeland security: A congressional guide for making America safer*. The Heritage Foundation. Retrieved April 7, 2005 from www.heritage.org/Research/HomelandDefense/bg1835.cfm

Chan, S. (2005a, August 23). M.T.A. turns to Lockheed for security. *The New York Times*, p. B1.

Chan, S. (2005b, August 24). New cameras to watch over subway system [Electronic version]. *The New York Times*.

Chan, S. (2005c, August 3). $100 million for a tunnel: What tunnel. *The New York Times*, p. B1.

Chertoff, M. (2005, March 16). *Remarks to the George Washington University Homeland security policy institute*.

Chittester, C. G., and Haimes, Y. Y. (2006). Risks of terrorism to information technology and to critical interdependent infrastructures. *The Berkeley Electronic Press*, 1(4). Retrieved November 17, 2006 from www.bepress.com/forum/vol1/iss4/402.

Committee on Appropriations. (2006, September 25). *Highlights: fy07 homeland security appropriations bill*. Retrieved December 8, 2006 from the United States House of Representatives via http://appropriations.house.gov/

Committee on Transportation and Infrastructure. (2004, May 5). *Railroad security* (Publication No. 95–129 PS). Washington, DC: U.S. Government Printing Office.

Department of Homeland Security. (2004a, February 2). *Department of homeland security announces fy 2005 budget in brief*. Retrieved March 8, 2004 from www.dhs.gov/

Department of Homeland Security. (2004b, March 22). *Rail and transit security initiatives*. Retrieved December 7, 2006 from www.dhs.gov/xnews/releases/press_release_0376.shtm

Department of Homeland Security. (2005a, July 13). *Secretary Michael Chertoff U.S. Department of Homeland Security second stage review remarks*. Retrieved December 7, 2006 from www.dhs.gov/xnews/speeches/speech_0255.shtm

Department of Homeland Security. (2005b, March 16). *Remarks for secretary Michael Chertoff U.S. Department of Homeland Security George Washington University Homeland Security Policy Institute*. Retrieved December 7, 2006 from www.dhs.gov/xnews/speeches/speech_0245.shtm

Department of Homeland Security. (2006a). *National infrastructure protection plan*. Retrieved December 8, 2006 from www.dhs.gov/xlibrary/assets/NIPP_Plan.pdf

Department of Homeland Security. (2006b, July 11). *Explosive detection technologies to be tested at Jersey City's exchange place station*. Retrieved December 7, 2006 from www.dhs.gov/xnews/releases/press_release_0948.shtm

Department of Homeland Security. (2006c, July 7). *Securing our nation's rail systems*. Retrieved December 7, 2006 from www.dhs.gov/xnews/releases/press_release_0944.shtm

Department of Homeland Security, and U.S. Department of Transportation. (2006, June 23). *Freight rail security action items*. Retrieved December 8, 2006 from the Transportation Security Administration via www.tsa.gov/what_we_do/layers/trip/freight_rail_security.shtm

Edmonson, R. G. (2006a, September 11). Five years later. *The Journal of Commerce*, 7(37), 10–15.

Edmonson, R. G. (2006b, September 18). Inside the port security act. *The Journal of Commerce*, 7(38), 12–15.

Elliott, A., and Lueck, T. J. (2005, July 25). Penn Station evacuated after bomb threat. *The New York Times*, p. B1.

Falkenrath, R. A. (2004). The 9/11 Commission report: A review essay. *International Security*, 29(3), 170–190.

Fiorill, J. (2005, May 17). U.S. seeks to improve hazardous rail shipment information flow, researches structure of tankers. *Global Security Newswire*. Retrieved May 5, 2005 from www.nti.org/d_newswire/issues/2005_5_17.html

Flynn, S. E. (2005, May 25). Color me scared. *The New York Times*, p. A25.

Gallagher, J. (2005, October 3). Stacking containers higher. *Traffic World*, 269(40), 23–26.

Grimm, C. M., and Sapienza, H. J. (1993, Spring). Determinants of shortline railroad performance. *Transportation Journal*, 32(3), 5–13.

Grosskopf, K. R. (2006). Evaluating the societal response to antiterrorism measures. *Journal of Homeland Security and Emergency Management*, 3(2), article 1. Retrieved December 8, 2006, from www.bepress.com/jhsem/vol3/iss2/1/

Hall, J. (2005, February 23). Paying the price for safety. *The New York Times*, p. A19.

Higham, S., and O'Harrow Jr., R. (2005, June 30). The high cost of a rush to security. *The Washington Post*, p. A01.

Hunton and Williams LLP. (2004, April). *Client alert: Rail Security Act of 2004*. Retrieved December 11, 2006 from www.hunton.com/files/tbl_s10News/FileUpload44/12601/GADM-0116_Rail_Trans69AE9.pdf

Johnston, V. R., and Plant, J. F. (2008). Rail security after 9/11: Toward effective collaborative regulation. *Public Works Management & Policy*, 13(1), 12–21.

Journal of Commerce (2006, October 23). *Rail Cargo Statistics*, 48.

Karmarck, E. C. (2002, November). Applying 21st-century government to the challenge of homeland security. *The Berkley Electronic Press*, 1(2). Retrieved November 17, 2006, from www.bepress.com/forum/vol1/iss2/art5

Kocieniewski, D. (2005, May 9). Facing the city, potential targets rely on a patchwork of security. *The New York Times*, p. A1.

Larsen, R. J. (2005, May 20). Years after 9/11. . . no one is in charge of fighting our two biggest threats. *The Washington Post*, p. A21.

Lee, H. L. (2004, September 3). *Supply chain security – are you ready* (Publication No. SGSCMF-W1–2004). Stanford, CA: Stanford Global Supply Chain Management Forum.

Lipton, E. (2005a, February 20). Audit faults U.S. for its spending on port defense. *The New York Times*, p. 1.1.

Lipton, E. (2005b, May 25). Loopholes seen in U.S. efforts to secure overseas ports. *The New York Times*, p. A6.

Lipton, E. (2005c, May 8). U.S. to spend billions more to alter security systems. *The New York Times*, p. 1.1.

Maltz, A. (1993, Spring). Private fleet use: A transaction cost model. *Transportation Journal*, 32(3), 46–53.

Martonosi, S. E., Ortiz, D. S., and Willis, H. H. (2005). Evaluating the viability of 100 per cent container inspection at America's ports. In H. W. Richardson, P. Gordon, and J. E. Elliot (Eds.), *The economic impacts of terrorist attacks* (pp. 218–241). Northampton, MA: Edward Elgar.

Metro-North Operating Statistics (2006). Retrieved January 6, 2007 from www.mta.org/no/mnr

Millar, F. (n.d.). *New strategies to protect America: Putting rail security on the right track.* Retrieved November 30, 2006 from www.americanprogress.org/

Mineta Transportation Institute. (2005, September 29). *Third national transportation security summit: Rail security – a symposium on terrorism and business continuity.* (MTI Report No. S-05–02). San Jose, CA: Author.

Morris, G. D. L. (1995, February 1). Railcar makers take a hand in reengineering chemical fleets. *Chemical Week*, 156(4), 32.

Morris, G. D. L. (2006). Chemical producers put new emphasis on logistics and security: Responsible care becomes the common ground for shippers and service providers. *Inbound Logistics*, 6(Suppl. 1), 60–64.

National Research Council. (2010). *Review of the department of homeland security's approach to risk analysis*. Washington, DC: The National Academies Press.

New York Susquehanna and Western Railroad Police. (2007). *Homepage information.* Retrieved January 6, 2007 from www.NYSWrailroad police.org.

Norfolk Southern. (2005). *In Katrina's wake: Norfolk Southern responds and recovers (DVD).* Norfolk, VA: Norfolk Southern Corporation.

Norfolk Southern Police. (2007). *History of railroad police*. Retrieved January 6, 2007 from http://NSPolice.com/History0.htm

Oberstar, J. L. (2005, May 12). *Introducing the Rail Security Act of 2005*. Retrieved December 7, 2006 from United States House of Representatives via www.house.gov/transportation_democrats/Rail%20Security%20Act%20(2005).pdf

Our unnecessary insecurity. (2005, February 20). *The New York Times*, p. 4.8.

Persico, J. E. (2001). *Roosevelt's Secret War: FDR and World War II Espionage*. New York: Random House.

Peterman, D. R. (2005, January 24). *Passenger rail security: Overview of issues* (Congressional Research Service Report for Congress No. RL32625). Washington, DC: Congressional Research Service.

Peterman, D. R. (2006, June 23). *Transportation security: Issues for the 109th congress* (Congressional Research Service Report for Congress No. IB10135). Washington, DC: Congressional Research Service.

Plant, J. F. (2007). Enhancing rail security. In J. F. Plant (Ed.), *Handbook of transportation policy and administration* (pp. 595–606). New York: CRC Press.

Plant, J. F., and Young, R. R. (2007). *Assessment of terrorist threats to the U.S. railway network.* Boston: Citizens for Rail Safety.

Rail and Public Transportation Security Act of 2005, H.R.153, 109d Cong. (2005).

Real security, or politics as usual: an insecure nation (2005, May 1). *The New York Times*, p. 4.13.

Redding Jr., R. R. (2005, April 21). Officials prepared for long hazmat suit [Electronic version]. *The Washington Times*.

Roberts, P. (2005, August). Rail-thin security. *CSO Magazine*. Retrieved December 7, 2006 from www.csoonline.com/read/080105/wonk_wonk.html

Rothberg, P. F. (2005, April 6). *Hazardous materials transportation security: Highway and rail modes* (Congressional Research Service Report for Congress No. RL32851). Washington, DC: Congressional Research Service.

Roucolle, G., and Lowes, R. (2006, Spring/Summer). Rail transit and the security challenge. *Mercer on Transportation & Logistics*, 12, 27–31.

Rugy, V. (2005, March 18). Homeland preparedness [Electronic version]. *The Washington Times*, S. 1379, 109th Cong.

Series of bombs explode on 7 trains in India, killing scores (2006, July 12). *The New York Times*, p. A1.

Stern, J. (1999). *The ultimate terrorists*. Cambridge, MA: Harvard University Press.

Supply Chain Council. (2006). *Supply chain operational reference model, version 8.0*. Pittsburgh, PA: Supply Chain Council.

Thompson, W. C. (2002, January 14). *Railroad infrastructure security*. Proceedings of the TRB Annual Meeting, Session 107.

(The) Trailer Train Story (undated). Chicago, IL: Trailer Train Corporation.

Transportation Security Administration. (2006, July 7). *Fact sheet: Securing our nation's rail systems*. Retrieved December 11, 2006 from www.tsa.gov/assets/pdf/RailSecurity FactSheet.pdf

Transportation Security Administration. (n.d.). *Our security strategy: Systems based perspective*. Retrieved December 7, 2006 from www.tsa.gov/approach/systems.shtm

TTX Corporation. (2007). *Background information*. Retrieved January 6, 2007 from www.TTX.com.

Transportation Research Board. (2005). *Critical issues in transportation*. Washington, DC: Transportation Research Board.

U.S. appellate court stays District of Columbia hazmat ban [Electronic version]. (2005, April 19). *CSX*.

U.S. Department of Transportation. (n.d.). *Section-by-section analysis of the Federal Railroad Safety Enhancement Act of 1999*. Retrieved December 11, 2006 from http://testimony.ost.dot.gov/final/secfra.htm

U.S. Government Accountability Office. (2005a, October 20). *Passenger rail security: Enhanced federal leadership needed to prioritize and guide security efforts*. GAO-06–181T. Retrieved February 22, 2006 from Government Accountability Office via www.gao.gov/htext/d06181t.html

U.S. Government Accountability Office. (2005b, September). *Passenger rail security: Enhanced federal leadership needed to prioritize and guide security efforts*. GAO-05–851. Retrieved February 22, 2006 from Government Accountability Office via www.gao.gov/htext/d05851.html

U.S. Government Accountability Office. (2006a, June 21). *Freight railroads: Preliminary observations on rates, competition, and capacity issues*. GAO-06–898T. Retrieved December 7, 2006 from Government Accountability Office via www.gao.gov/htext/d06898t.html

U.S. Government Accountability Office. (2006b, March 29). *Passenger rail security: Evaluating foreign security practices and risk can help guide security efforts*. GAO-06–557T. Retrieved December 7, 2006 from Government Accountability Office via www.gao.gov/htext/d06557t.html

United States Senate Committee on Commerce, and Science & Transportation. (n.d.). *Summary: Transportation Security Improvement Act of 2005*. Retrieved December 11, 2006 from http://commerce.senate.gov/pdf/Summary.TSIA%20Intro.pdf

The USA Patriot Act. (n.d.). Retrieved October 26, 2004 from www.epic.org/privacy/terrorism/usapatriot/

Wald, M. L. (2004a, June 30). Amtrak chief says finances are better, delays worse. *The New York Times*, p. A21.

Wald, M. L. (2004b, November 27). Reports faults F.A.A action on handling risky cargoes. *The New York Times*, p. A10.

Young, R. R., Swan, P. F., and Burn, R. (2002, Winter). Private railcar fleets: The customer return problem. *Transportation Journal*, 42(1), 51–61.

Young, R. R., and Stenger, A. J. (2001, Winter). Maintenance practices among private railcar fleet operators in the chemical and plastics industries. *Transportation Journal*, 41(1), 61–70.

Yuva, J. (2005, November). How secure is your supply chain? *Inside Supply Management*, 16(11), 24–31.

2 U.S. railroad systems overview

The U.S. rail system is a complex structure for myriad reasons. While operating as a system, it is comprised of a mosaic of operators that include corporations and government agencies; services that encompass freight, intercity passenger, and regional and commuter passenger; connections to railroads serving Canada and Mexico; and connections with other modes of transportation including trucking and maritime.

An individual railroad has its own network consisting of terminals, rail yards, junction points, and sidings with customers and consignees all connected with lines comprising the links in the system. Moreover, all of the interconnecting railroads in North America represent a network system whereby both rolling stock and locomotives are engaged in extensive interchange so that shippers and consignees in the wider trading bloc market are served in a timely and efficient manner.

To better understand rail operations, the topic needs to be segmented into the headings of *operating entities*, *freight versus passenger operations*, *infrastructure*, and *rolling stock*. For the purposes of discussion, this book intentionally excludes light rail passenger systems, narrow gauge railroads, and specialty tourist operations because those systems generally do not regularly connect with the general railroad system of transportation[1] despite often operating their tracks in adjacent rights-of-way of the larger network.

Finally, since the lexicon of railroad transportation is unique unto itself, a detailed *glossary* is provided to assist the reader in achieving a working understanding of this mode that is now approaching its 200th anniversary.

Operating entities

The U.S. rail system consists of more than 5550 railroads of which greater than 95 percent are freight carriers segmented into what are referred to as Class I, or railroads having more than approximately $433 million annual revenues; regional or Class II railroads, smaller than Class I railroads but having more than $40 million annual revenues and/or more than 350 miles of track; local linehaul shortline or Class III railroads, which are smaller than the regional railroads by either measure; and switching and terminal railroads that primarily provide interchange services within very narrowly defined geographical limits. Collectively, these represent more than 168,755 track miles[2] (AAR 2013). An approximate breakdown is provided in Table 2.1.

The legal organization of the largest railroads is that of a publicly traded corporation. Six of the seven Class I railroads trade on the New York Stock Exchange; however, the seventh, the Burlington Northern Santa Fe, is now a unit of Berkshire Hathaway Corporation, which is a publicly traded conglomerate. Many of the smaller railroads are privately owned by online captive industries that are basic in businesses other than transportation, such as mining, forest products, or electricity generation. A notable development over the past 25 years has been the establishment of firms holding a portfolio of smaller (non–Class I) railroads, the two largest examples being RailAmerica Incorporated and Genesee & Wyoming Incorporated (G&W), with RailAmerica merging with G&W in 2013.

Having a higher public profile, but being significantly smaller in size, are the passenger railroads consisting of a single federally controlled intercity carrier, the National Passenger Rail Corporation, better known as Amtrak; a formerly state-controlled passenger and freight carrier, the Alaska Railroad; and the regional commuter rail authorities that either are publicly owned and operated or are operated as a franchise to a regional transportation authority that owns the tracks. A summary of the organizations operating this segment is found in Table 2.2.

Table 2.1 Classes of U.S. railroads

Freight railroad classification	*Number of firms*	*Miles of road*[a]
Class I	7[b]	95,235
Class II or regional[c]	31	16,073
Linehaul shortline[c]	309	20,155
Switching and terminal[c]	205	7,014
Total	552	138,477

Data source: Association of American Railroads

Notes

a Miles of track includes trackage rights of one railroad over the right-of-way of other railroads.

b Includes the five largest U.S. railroads plus the two Canadian railroads whose respective presence in the U.S. would qualify each for Class I status.

c Calculated from various sources.

Table 2.2 Major U.S. passenger rail operations

Organization	*States of operation*	*Miles of track*
Amtrak	Nationwide	730[a]
Alaska Railroad	Alaska	470 miles
Sum of U.S. commuter operations	Multiple[b]	~5,000

Data sources: various agency websites

Notes

a Amtrak-operated lines, increases to over 22,000 miles when trackage rights are included.

b Details of those systems are provided in Chapter 6.

Freight versus passenger operations

Freight

In selling railroad services, the freight carriers work in car units or train units. Most shippers may consign one or several carloads[3] to a given customer; there are also situations where railroads sell the services of an entire train to a single customer. This occurs in two principal situations: unit trains and intermodal trains. Unit trains are where an entire train, usually hauling a single commodity, such as coal or grain, is moved from a point of origin to one destination, such as from a mine to a power plant or steel mill. Intermodal capacity is sold either as an entire train to a customer, such as UPS, or on a wholesale basis to an intermodal agent, such as Hub Group, Pacer Stacktrain, or Alliance Shippers. The character of the intermodal relationship is important because in terms of knowing what is being carried, the railroad identifies only a loaded or empty container or trailer; hence, with the exception of hazmat, it has little knowledge of the exact contents.

From a historical standpoint, freight traffic, as previously discussed, has been increasing steadily ever since the railroads were economically deregulated with the passage of the Staggers Act of 1980. Since that time they have been permitted to sell off low-volume lines to shortlines and regional railroad operators, an event that paved the way for the current configuration of the railway network. Additionally, they were permitted to abandon lines that had such low volume that they could neither be sold off nor operated economically (refer to Table 2.1, which shows the number of non–Class I railroads). Many railroads also seized this opportunity to reduce infrastructure investment by single tracking what were, at one time, double-track lines, double-tracking former triple-track lines, and abandoning some rail yards completely (Coyle and Langley 2006, pp. 129–139).

Typically, the railroads have been characterized as being highly capable of moving large quantities of materials and goods long distances. With the development of improved highway systems in the U.S., much of the freight rail traffic has focused on bulk materials such as chemicals, plastic resin, coal, coiled steel, forest products (including paper and lumber), and compressed gas, such as propane and LPG (liquefied petroleum gas). Large volumes of manufactured goods such as finished automobiles as well as their component parts are also important rail-borne freight. Table 2.3 shows the ton-miles[4] by sector.

Intermodalism

To end the consideration of the rail network at the industry level, however, is to short-change discussion of the topic. In U.S. history, three events converged to provide the basis for the current intermodal construct that represents the industry today. The first was in the mid-1800s, when the concept of carrying movable wagons on a flatcar was pioneered by the Camden and Amboy, a predecessor of the Pennsylvania Railroad – a development that ultimately evolved into the movement of highway trailers and what was termed *piggyback* (Cunningham 1951).

Table 2.3 Ton-miles (mio) of freight by rail for 2015

Commodity	*Ton-miles*	*Commodity*	*Ton-miles*
Coal	683	Metals	46
Chemicals & allied prod.	180	Stone, clay & glass	43
Farm products	145	Waste & scrap materials	37
Non-metallic minerals	160	Petroleum & coke	50
Intermodal	120	Forest products & paper	60
Food & related products	101	Metallic ores	59
Motor vehicles & equip.	23	Other	24
Total ton-miles			1,731

Data source: U.S. Bureau of Transportation Statistics

The second event was Malcolm McLean's[5] development of containerized ocean shipping – now ubiquitous for the movement of packaged goods internationally. The third was the Stagger's Rail Act that economically deregulated the industry and moved trailer-on-flatcar (TOFC) and container-on-flatcar (COFC) traffic from common carriage to contract carriage[6] status. These developments demand that any study of rail freight traffic include both trucking and ocean shipping – rail's partners in intermodal commerce (Coyle 2006, pp. 209–214).

During the past 30 years, the most consistent growth has occurred in the intermodal sector, which began as TOFC to become the ubiquitous ocean as well as domestic containers (COFC) now transported on specialized railcars often now with the ubiquitous double-stack units. As shown in Table 2.4, the 35-year period since economic deregulation marks a period of growth in intermodal traffic that can be attributed to both an improvement in railroad performance and an increase in a range of problems plaguing the motor carriers.

Railroad performance has improved as rail carriers have made massive investments in rights-of-way that include the laying of continuously welded rail (CWR), newly ballasted track, super-elevated curves, and efficient COFC/TOFC terminals. Conversely, the trucking industry has been plagued by a plethora of issues such as driver shortages, hours of service regulation that further constrain driver availability, increasing traffic congestion as urban areas continued their outward sprawl, and the increasing number of vehicles on the nation's roadways. Consequently, the railroads' share of total ton-miles in the U.S. has been slowly increasing and now stands close to 40 percent.

These developments are significant to the discussion of terrorism and infrastructure protection (1) because of the sheer volume of ocean containers or truck trailers being hauled, both as loaded and empty, and (2) because containers and trailers are far more mobile than railroad cars since they ultimately are taken off the rails and moved over roadways.

Decades ago, the railroads made a conscious effort to exit the retail marketing of intermodal services with the result being that today the railroads sell container positions or trailer positions on flatcars or double-stack cars to the major trucking companies such as J. B. Hunt, Schneider National, and UPS and to the ocean container lines such as Maersk, Evergreen, and Hapag-Lloyd.

Table 2.4 Growth in intermodal volume since deregulation

Year	*1980*	*1985*	*1990*	*1995*	*2000*	*2005*	*2010*
Millions of trailers and containers	3.0	4.6	6.1	8.0	9.1	11.7[a]	11.3
Growth rate (year over year)		53.3%	32.6%	31.1%	13.8%	28.8%	−3.4%

Data source: Association of American Railroads

Note

a Approximately 75 percent of the most recently reported volume is domestic and ocean containers.

The railroads interviewed for this research stated that they specifically exclude hazardous materials from intermodal service. To succeed at this, they need to rely on the systems and procedures of the truckers, ocean carriers, and intermodal retailers. Ostensibly, the most obvious way to detect hazardous materials is by the presence of the DOT-mandated placards placed on the exterior of vehicles. Another innovation in the past decades is the tank container, otherwise known as an ISO[7] tank, which is a bulk vessel suspended in a steel framework and handled in the same manner as 20-foot ocean containers when loaded aboard ship or mounted on highway chasses for over-the-road travel. ISO tanks carrying hazardous materials are also prohibited from railroad intermodal service; however, when seen on freight trains, they are likely hauling non-hazardous materials, including bulk beverages, food items, and numerous industrial products.

The intermodal flatcars and double-stack equipment are not owned by the railroads per se but may be owned by the ocean container carriers such as Maersk and American President Lines; trucking companies, such as Schneider National and J.B. Hunt; or intermodal services firms, including Pacer Stacktrain, the Hub Group, and Alliance Shippers. Most often the equipment is owned by TTX Corporation (formerly Trailer Train Corporation), a separate legal entity owned by a consortium of railroads, but acting independently of them (*Trailer Train*, undated).

Passenger traffic

In recent years the volume of passenger rail traffic has been making a comeback as well. While Amtrak has received and continues to receive the majority of press coverage, especially with regard to ongoing congressional funding discussions, the growth in interest in commuter rail is especially noteworthy. Commuter services were originally the passenger operations of both large and small railroads alike in key urban markets such as Boston, New York, Chicago, Washington, as well as others. It is now nearly the exclusive domain of regional transportation authorities[8] that include the Massachusetts Bay Transportation Authority (MBTA), NJ Transit Rail, Southeast Pennsylvania Transportation Authority (SEPTA), and others. Commuter rail operations typically have most of their operations within approximately a 100-mile radius of their respective city centers. Growth is attributed to a combination of urban sprawl, highway traffic congestion, fuel prices, and environmental considerations such as air-quality standards. Table 2.5 identifies many of the heavy rail systems in the U.S. and their current ridership. Note that

Table 2.5 Average daily commuter ridership in 2010

Commuter rail entity	*Metropolitan areas served*	*Date opened* *Former owner*	*Passengers/day (thousands)*
Caltrain	San Francisco to San Juan Corridor	1987 Southern Pacific[a]	35.6
Long Island RR (MTA)	New York City	1836 Pennsylvania[b]	354.1
Maryland Area Regional Commuter RR	Baltimore and Washington, DC	1984 Chessie[c]	30.0
Metropolitan Boston Transportation Authority	Boston	1973 Boston & Maine[d]	131.4
METRA (Chicago)	Chicago	1984 various	301.6
Metrolink	Los Angeles	1992	39.7
Metro North Commuter RR (MTA)	New York City	1983 Conrail[e]	265.0
NJTransit Rail	New York City and Philadelphia	1983 Conrail	291.4
Southeast Pennsylvania Transportation Authority	Philadelphia	1983 Conrail	124.4
Virginia Railway Express	Washington, DC	1992	17.5
South Florida Tri-Rail	Miami	1987	12.5
Northern Illinois Commuter	Chicago	1903	11.9
Sounder Commuter	Seattle & Tacoma	2000	9.1
Trinity Railway Express	Dallas/Ft. Worth	1996	8.8
Utah Transportation Authority	Salt Lake City	2008	5.1
Capital Corridor	Sacramento & San Jose	1998	4.8
NCTD Coaster	San Diego	1995	4.3
New Mexico Rail Runner	Albuquerque	2006	3.8
Altamont Commuter Express	San Jose	1998	2.5
Shore Line East	New Haven	1990	2.0
Northstar Line	Minneapolis	2010	1.9
Westside Express	Beaverton	2010	1.2
Music City Star	Nashville	2006	0.9
Capital MetroRail	Austin	2010	0.7
TOTAL			1,661.2

Data sources: various agency websites

Notes

a Predecessor is now a part of Union Pacific.

b Predecessor later merged into Conrail and subsequently split between CSX and Norfolk Southern.

c Predecessor became CSX.

d Predecessor became part of Guilford and now known as Pan Am Railways.

e Subsequently split up between CSX and Norfolk Southern.

while many of these commuter agencies also operate light rail services, these elements are excluded from any detailed discussion in this book.

In many instances, freight and passenger carriers share significant miles of line. While Amtrak owns the heavily used route between Washington, DC, and Boston as well as the Keystone Line between Philadelphia and Harrisburg, portions of those lines also carry commuter rail traffic, including Maryland Area Rail Commuter (MARC), SEPTA, Metropolitan Transportation Authority (Long Island Railroad and Metro-North in New York and Connecticut), New Jersey Transit (NJT), and MBTA. Chicago's METRA and California's CALTRANS are other examples of major commuter operations sharing trackage with freight railroads (Carstens 1985; Walker 1994).

Additionally, Amtrak and regional commuter lines may have freight traffic operating over their lines using trackage rights.[9] Conversely, Amtrak runs on trackage rights over freight railroads in areas other than the Northeast Corridor and the Keystone Line. Some commuter operations – such as Maryland DOT's service over CSX to Brunswick, MD, and Martinsburg, WV, and the Virginia Railway Express in northern Virginia – operate over the freight railroads. Moreover, there are many occasions where passenger railroads and freight railroads parallel each other in close proximity even to the extent of having their respective tracks occupy the same rights-of-way in what many land-use planners refer to as transportation corridors. Examples are the close proximity of CSX's mainline between Washington and Pittsburgh with the lines of the Washington Metro system in Maryland; and Norfolk Southern's line paralleling Amtrak's Keystone Line for several miles south of Harrisburg, PA.

The volume of passengers handled by Amtrak and the various commuter agencies begins to approach the volume handled by the nation's domestic air carriers. The primary difference, however, is the nature of the network. While average trip distances for rail are significantly shorter than those of commercial aviation, the number of station locations is much greater. Any comparison to air travel is instructive as the smallest regional airports will accept aircraft with as few as 19 passengers (e.g. Beech 1900 and BAe 31 aircraft) where capacity utilization may require a yield factor of 60 percent, or 12 passengers, to achieve economic viability. By comparison, many commuter trains will make station stops for as few as one passenger.

A further distinction between passenger rail service and commuter airlines is that the latter typically operate in a hub-and-spoke environment where an entire flight is both loaded and offloaded en mass. Conversely, a commuter train may have as many as 10–20 stops with extensive loading and offloading of passengers at just a few locations (NJ Transit 2006; Metro-North 2006). While Amtrak provides daily commuter operations in several major metropolitan areas, the role played by the regional authorities remains significant and a major consideration within this research. A summary of Amtrak and commuter rail operations was provided in Table 2.5.

Amtrak

The National Passenger Rail Corporation, or Amtrak as it is better known, was established by Congress in 1970 with the specific purpose of relieving the major railroads of their money-losing passenger operations while still retaining passenger rail service for the nation. It is, therefore, no accident that Amtrak acquired track and other infrastructure that (1) remains in close proximity to freight rail operations and (2) still utilizes the tracks of freight railroads through much of its network.

Amtrak's densest route, the Northeast Corridor, daily moves large volumes of passengers between a limited number of key nodes, namely the District of Columbia, Baltimore, Philadelphia, Metropark (Edison, NJ), Newark, New York, New Haven, Providence, and Boston. Although not specifically a commuter railroad, many of Amtrak's passengers are, indeed, commuters. This is true of the Northeast Corridor, Pennsylvania's Keystone Service, and those trains operated for California's Caltrain. The railroad operates hundreds of trains per day, moving an average of 750,000 passengers (Amtrak 2006). Not unlike the airlines with additional flights, Amtrak will accommodate additional passengers by either adding extra cars to trains or by providing extra trains during peak holidays.

Infrastructure

The U.S. rail network consists of approximately 168,000[10] miles of track that, if double track, traverses 66-foot wide rights-of-way. This average accounts for single-line routes that are narrower as well as wider multi-line routes and rail yards. Statistics show miles of track and miles of railroad with the latter designating rights-of-way with multiple tracks (AAR 2013). Exclusive of yards and terminals, the resulting total land mass controlled nationally is roughly the equivalent of 3,500 square miles or an area that approaches the size of the State of Delaware.

The rail network needs to be considered as a system for myriad reasons. First, it is operated by several hundred private firms, the majority of which connect with one or more others to which must be added the government-owned passenger carriers that include Amtrak and the commuter lines. Second, because of the use of standard gauge, the rail networks of Canada and Mexico also have multiple connections with the U.S. Third, because the economic realization that railroads are typically most adept at moving large quantities of goods long distances. Large quantities of goods can be characterized as bulk materials, dimensional goods (that the railroads will frequently term *heavy, high, and wide loads*), and containerized traffic where packaged goods are aggregated into containers that are typically 8′ × 8′ × 20′; 8′ × 8′ × 40′; or 8′ × 8′ × 53′.

From an analytical standpoint, various sections of this book will make reference to *links* and *nodes*. This is typically a means whereby logisticians and transportation policymakers evaluate transportation networks with *nodes* making reference to fixed points of geography such as stations, terminals, and interlockings. With freight transportation, this can also make reference to both supplier and customer locations. Conversely, *links* refers to transportation lines that connect those nodes. Obviously, the more links and nodes in a particular transportation system, the more complex it will be, especially when more than two links connect to a single node. It is also worthwhile to note that individual nodes may serve as an origin or destination for a great many links, an example being Norfolk

Southern's Pittsburgh Conway yard, where trains are received and dispatched to such key nodes as Cleveland, Harrisburg, and Indianapolis. Hence, the more connecting links, the greater importance that such a node would have for the system (Coyle and Langley 2003, p. 55).

Railroads need to be viewed not only in a total transportation perspective but as enablers in the supply chain. Although it is a term that has now reached common parlance, it is instructive to define the term *supply chain*. Perhaps the most useful explanation can be found in the Supply Chain Operations Reference (or SCOR) Model developed and furthered by the Supply Chain Council whereby (1) a chain is a group of buying and selling members that are dependent upon one another, (2) each entity has the business activities of *source*, *make*, and *deliver* with an overarching or coordinating *plan* activity, and (3) three flows transcend all of the members of the supply chain – namely *physical*, *informational*, and *financial* flows (Supply Chain Council 2006). Transportation is significantly present at the inter-firm connections where one firm's *source* is the *deliver* function of another.

The international economy is dependent upon supply chains, but suffice it to say that every economy consists of a myriad of combinations comprising millions of supply chains. Disruption of an economy is a simple and straightforward matter: interruptions of its supply chains insofar as its physical flows are impacted. Most people think only of terrorist attacks on the physical flows because that is what is visible. From a terrorism standpoint, that is where the most newsworthy events may occur, but consider that the physical flow is significantly enabled by the information flow. This is to say that an attack on the information technology capabilities of transportation providers will disable an economy perhaps with greater effect than any attack on the physical flow, specifically infrastructure (Meade and Molander,2006; Lee 2004).

Infrastructure consists of rights-of-way – those corridors where track can be found plus bridges, tunnels, trestles, culverts, classification yards, engine and car repair facilities, intermodal terminals, passenger stations, control towers, data centers, and office buildings. The most visible are the terminals, stations, bridges, and tunnels, but, from a terrorist standpoint, all represent potential targets depending upon the type of threat that can be considered. Physical protection of all of this capital investment is a difficult task given that the seven Class I railroads plus the aggregate of the shortline and regional railroads are estimated to have in excess of 100,000 bridges and culverts.

Communications and signaling

For railroads, the information flow is comprised of transactional systems just the same as any business or government agency. However, railroads have extensive communications and signaling systems that control the flow and speed of trains across their respective networks. All entities dealing in goods – whether they are manufacturing, selling, or transporting – need to be able to answer one fundamental question: *where is it?* The inability of manufacturers to answer this question with regard to raw materials – typically done with inventory and warehouse management systems – ultimately shuts down production. Similarly, should firms (whether producers, wholesalers, or retailers) incur such inability, stock-outs or shortages are likely to occur for consumer products. For transporters, the inability to answer the question of *where is it?* results in the inability to determine which rail cars to move where. All of these scenarios, however, carry highly damaging consequences, including disruption to financial systems, as buyers and sellers would no longer have the ability to engage in commerce.

Terminals

Protection of intermodal terminals, a considerably smaller number nationwide, also poses a very significant challenge even though the current number of terminals in the U.S. is only total of 160. These nodes may or may not be located on the owning railroad and many may be shared with intermodal services firms. Note that the transnational character of several firms requires that terminals located in Canada and Mexico be included.

Passenger operations have stations numbering in the hundreds, some at locations boarding thousands of passengers per day; others, as few as a dozen. What is germane is that the rail system is a network having some characteristics much in common with the airline system – someone entering the system at a small, unprotected station is *still in the system*.

Urban versus rural network elements

Economic development and railroad service have been mutually dependent since the founding of the first railroad, the Baltimore and Ohio in 1830. Railroad companies built lines to access markets; however, clusters of communities were created to access "improved transportation" provided by the railroads. This provides a challenging conundrum with regard to security: how to retain railroad service to population centers already challenged by a labyrinth of overused roadways while lowering any perceived threat that the railroads may pose because they may be attractive targets for terrorist activity or a useful means for executing terrorist missions (TRB 2005, p. 3).

Urban routes also tend to have freight and passenger trains either running over the same tracks or running on separate tracks in close proximity to one another. There are many examples of both. In the greater New York area, the Long Island Railroad (MTA) moves passengers over the same tracks as the freight carrier New York and Atlantic Railroad; SEPTA shares various tracks with Norfolk Southern in the Philadelphia area. Throughout the Northeast, Amtrak, PATH (Port Authority Trans-Hudson), NJ Transit, Norfolk Southern, Conrail Shared Assets, and various shortlines – including Morristown and Erie, and New York, Susquehanna, and Western (SusieQ) – operate rails through common transportation corridors (Carstens 1985; Walker 1994). Such corridors are, at least in part, an efficient approach to rationalizing land use and often include other utilities, including water lines, petroleum pipelines, electricity transmission, and telecommunications.[11]

In contrast, there are thousands of miles of rail line that pass through rural areas, some marked by small towns and villages; others, by long miles of open spaces. The common theme shared by both urban and rural areas is how to respectively protect the infrastructure and residents of each. Urban areas with higher densities of rail routes may offer alternative routes should one be taken from services; however, even these have their critical "choke" points. Rural areas may offer fewer alternative routings, but they have the advantage of more difficult access, both from a geographic standpoint and a social one, where strangers may be more likely noticed.

Interchange

The U.S. railroad network[12] operates on a standard track gauge of 4′ 8½″ that allows the rolling stock of one railroad to operate over the rails of another provided that they have connection points. There are some railroads that do not connect with others and are captive to a particular industry, most of which are employed by mining and forest product-firms for their own internal transportation. These will remain outside of the scope of our interest.

Interchange represents an inherently defensive capability for one railroad to route traffic around a particular problem or sensitive area. Historically, interchange provides the opportunity for one railroad to use a competitor's tracks to route around derailments or storm damage.

NAFTA connections

Large volumes of freight move north and south between the U.S., Canada, and Mexico. All three countries use the same standard gauge and interchangeable equipment. The volume involved poses its own challenge as goods must clear U.S. Customs' formalities upon entry. Although goods are administratively processed for entry into the U.S., intensive inspections, if any are being performed, occur at rail yards some distance from the borders, although Customs and Border Protection operate x-ray and radiation detectors at many ports of entry or just inside the border. Physical inspection of every railcar does not occur as bill of lading information regarding shipper, consignee, cargo description, and cargo quantity are profiled to determine appropriate levels of scrutiny.

It is noteworthy that many containers arriving at Halifax, Montreal, Prince Rupert, and Vancouver may have a U.S. consignee as the final destination. These are unloaded from vessels at the Canadian ports and transported under bond (note that these may be marked TIR for *Transports Internationaux Routiers*) to a designated U.S. port of entry for clearance. Ports of entry are not necessarily that port closest to the border – note that many containers arriving at Montreal and Halifax clear customs at Chicago rail yards.

Should the Mexican ports of Ensenada and Los Moches ever be fully developed as an alternative for U.S. West Coast ports, a similar mechanism can be envisaged. Moreover, with constrained U.S. port capacities, the Canadian port at

Prince Rupert, British Columbia, is already being developed with the outcome being more ocean containers passing through Canada en route to U.S. customers – it has the strategic advantage of being that North American deep-water port closest to Asia.

Rolling stock

Railcars and locomotives, also referred to as *rolling stock*, can be either owned or operated by the railroads or by non-railroad companies referred to as private fleet operators. The differentiator is asset utilization; hence, freight requiring specialized equipment has significantly lower utilization factors or annual equipment turns, meaning that railroads over time increasingly refused to make such investments.

Railroads own or operate a range of equipment. Passenger operations can be bifurcated into short-distance and long-distance activities, but this is also the major equipment differentiator. Longer-distance equipment is typified by higher levels of comfort, often with dining and sleeping capabilities, while shorter-distance equipment is focused on high-density seating and coaches intended for rapid boarding and detraining.

Freight railroads operate equipment generally configured for moving large volumes of cargo over long distances. Equipment is intended for bulk materials, oversized machinery, and intermodal operations.

Privately owned railcar fleets

From an economic standpoint, equipment ownership decisions focus on returns on investment that can be translated into maximizing the number of car trips per year. Finally, there are the occasional instances where locomotives are owned and operated by firms other than the railroads – most often this may involve an electric generating utility as was the case previously with Detroit Edison.

Much of the freight car traffic hauled by the railroads is owned privately and not by the railroads themselves. This is especially true when the goods being carried require specialized equipment or where a shipper's industry practice necessitates the use of cars for long-term storage. Both eventualities serve to reduce individual asset utilization to the point where the railroads would find respective ownership uneconomic.

Specialized equipment may mean tanks of specific alloys, unique lining systems, or having pressure ratings for compressed gases. Covered hoppers have low asset utilization factors because of their frequent use for product storage by either producers or their customers. The AAR reports that railroads operate 477,000 cars; industry shippers, 806,000 cars – 62 percent are therefore private (AAR 2013).

When the railroads refused to make the necessary investment in specialized equipment, the shipper did so with the result being the extensive car fleets owned and/or operated by individual firms as seen today. Sizeable fleets of tank cars are operated by such chemical makers as BASF, Dow, DuPont, and

Koch; and covered hoppers, by plastics producers, Shell, BP, Fina, and Huntsman. Agricultural products companies are also operators of tank cars and covered hoppers with such names as A. E. Staley, Corn Products, Cargill, and Archer-Daniels-Midland commonly found (Railway Equipment Register 2003).

Current constrained capacity

The existing transportation capacities of all modes are challenged making the economic impact of either a natural or manmade disaster potentially greater. Specifically, the concern is that other modes may have difficulty in making up for a loss of any specific piece of infrastructure. Where this argument may be effectively countered, however, may be with the demonstrated ability of Norfolk Southern being able to restore the Lake Pontchartrain, LA, bridge within just 12 days after its near total destruction by Hurricane Katrina (Norfolk Southern 2005).

Railroad capacity has been nearly constant since the early 1980s, when, following the passage of the Stagger's Rail Act, many miles of triple-tracked railroad were double tracked and many miles of double-tracked lines were single tracked. Clearly, it was a rational response to the steadily declining traffic of the 1950s, 1960s, and 1970s, but it left a situation where 30 years of growth equal only to the growth of gross domestic product at a compounded rate could be expected to double the traffic. Even in the mid-1990s, there were parts of the rail network that were suffering traffic congestion during some hours of certain days of the week.

Trucking has suffered as well, with ongoing driver shortages forcing many carriers to increase rates, refuse to take loads, and divert longer-distance hauls to intermodal service. The driver shortage has prompted some truckers to seek to make acquisitions, not for added customers or for expanded geographic coverage, but to acquire the drivers. In part the use of intermodal has been to improve driver lifestyles. Long-distance travel, difficult for family life and for health habits, was a driving force behind J. B. Hunt's decision to increase purchases of containers, chassis, and non-sleeper power as they sought to keep drivers closer to home.

The use of trucks has gotten more expensive as rates have been rising along with fuel surcharges to compensate for higher diesel fuel costs. Shippers, accustomed to 30 years of steadily decreasing rates, have retired many of their traffic executives having the institutional knowledge of when purchasing freight was a seller's market and have continued to expect continuing reductions in transportation rates. The result has been that many have been disappointed, with some shippers even having carriers refuse to do business with them.

Road congestion is now legendary in many regions of the country, where former traffic jams only during rush hour have been replaced by nearly round-the-clock congestion. As states and communities build new roads, they are only finding that new traffic patterns result as those capacities are quickly filling. Urban areas are clearly worse than rural, but congestion can now be found to be nearly ubiquitous (TRB 2005, p. 2; Journal of Commerce 2006; FHWA 2002).

Ports – especially West Coast load centers such as Los Angeles–Long Beach, San Francisco-Oakland, and Seattle – are capacity challenged on several fronts, including berth space, container yard acreage, and access roadways and rail lines. One example is Walmart, a major importer from Asia, which made a conscious decision to divert some shipments to Houston, a lower-utilized port, which requires goods to spend more time at sea as well as traversing the Panama Canal (TRB 2005).

Summary

The U.S. railroad network has undergone significant changes in the past 30 years. For freight railroads it can be reduced to several key points: (1) the dramatic rise in the movement of intermodal traffic as ocean containers, domestic containers, and trailers on flatcar; (2) a consistent rise in the volume of freight as railroads have improved their on-time performance and as the highway system has become increasingly congested; (3) the growth in coal and crude oil shipments (however, as this chapter is written, coal has seen a dramatic decrease due to the shuttering of many coal-fired power plants and the movement of crude oil stemmed by the drop in oil prices making imported oil a more cost-effective option); and (4) a steady increase in the use of passenger rail services, especially for commuter systems where there has been a significant increase in the number of systems within the past 10 years.

The system can be characterized by a level of complexity brought about by the large number of participating entities that allow the North American network to operate as a system. Nevertheless, those participating entities have different roles and different objectives given their differing customer industries and geographic focus, where it often becomes a situation of friendly cooperation juxtaposed with a modicum of intramodal competition. With regard to ownership of equipment, it is a matter of that owned by the various railroads that is typically operated in interchange, but also the railcar fleets owned or controlled by the non-rail firms, including chemical, electricity generation, and agricultural-product firms.

Passenger and freight systems co-exist perhaps often with relationships that can swing between cooperation and animosity. Even when not sharing each other's rails, these substantially different types of operations may have rails occupying common rights-of-way. Finally, there is the matter of trackage rights where entire trains of one railroad may operate over the rails of another.

Given all of these complexities, the challenges of protecting infrastructure are then reflected in (1) the overlapping property issues represented by the myriad railroads, (2) the overlapping legal responsibilities of law enforcement jurisdictions at the local, state, tribal, and federal levels, (3) the matter of private versus public sector interests as well as regulatory mandates given the ownership of the rights-of-way, (4) the public perception that railroads are by and large an antiquated and obsolete mode of transportation, and (5) the fact that the capacity of the U.S. infrastructure has come under increasing capacity constraints. All of these factors contribute to the vulnerability of the total network, a topic that will further unfold in the ensuing chapters. Several later chapters will provide detailed coverage of the movement of hazmat, a timely and complex topic that is germane to the overall discussion.

Notes

1 The general railroad system of transportation, as defined in 49 CFR 1580, Rail Transportation Security, refers to the network of standard gauge track over which goods and passengers are transported between cities and within metropolitan and suburban areas. Also refer to 49 CFR Part 209, Appendix A.
2 Track miles will differ from road miles. The latter excludes rail yards, sidings, and multiple tracks in a single right-of-way.
3 Often referred to as a block or cut.
4 A ton-mile is a measure of production equivalent to one ton hauled one mile.
5 McLean headed a trucking company bearing his name and first used a converted collier to move truck bodies without their wheeled chassis to improve productivity. However, that first voyage was a domestic one between New Jersey and Texas.
6 Beginning with the Act to Regulate Commerce of 1887, the *common carrier* designation stipulated that the railroads were prohibited from discriminating against particular shippers and had to carry the goods of all who demanded the service. *Contract carriage* means that particular rates and service attributes pertain to a specific contractual shipper in return for agreeing to a volume of freight between stipulated points.
7 ISO denotes a standard developed by the International Standards Organization, which is largely European.
8 There are new private initiatives underway in Florida and Texas, among other locations.
9 The practice of one railroad operating over the rails of another in exchange for payment which is tantamount to a toll fee.
10 Miles of track differs from miles of road. The latter excludes yards, terminals, and lines of double- or triple-track rights-of-way.
11 Note that Sprint was created in the nineteenth century and is an acronym for the Southern Pacific Internal Network Telephony.
12 There are some narrow gauge tourist and industrial operations, but they are captive and not considered part of the network.

Bibliography

Association of American Railroads (AAR) (2013). *Class I railroad statistics*. Retrieved May 12, from www.aar.org/StatisticsAndPublications/Documents/AAR-Stats-2013-02-07.pdf

Amtrak Operating Report. (2006). Retrieved January 6, 2007 from www.amtrak.com/servlet/contentserver/pagename=amtrak/am2copy/title_merge_copy_page&=copy&cid=108144267447&ssid=322.htp/october 2006.pdf

Carstens Publications (1985). *Handy Railroad Atlas of the U.S.* New York: Rand-McNally.

Cunningham, R. (1951). *Railroading in New Jersey*. Newark: Associated Railroads of New Jersey.

Federal Highway Administration. (2002). *Freight analysis framework: North American interchange on transportation statistics*. Washington, DC: U.S. Department of Transportation.

Genesee and Wyoming Incorporated (2013). *Annual Report 2012*. Available at http://ir.gwrr.com/annual-reports

Darien, Connecticut Journal of Commerce (2006, October 23). Rail Cargo Statistics, 48.

Lee, H. L. (2004, September 3). *Supply chain security – are you ready* (Publication No. SGSCMF-W1–2004). Stanford, CA: Stanford Global Supply Chain Management Forum.

Norfolk Southern. (2005). *In Katrina's wake: Norfolk Southern responds and recovers (DVD)*. Norfolk, VA: Norfolk Southern Corporation.

(The) Official Railway Equipment Register (2003). East Windsor, NJ: R.E.R. Publishing Corp.

Supply Chain Council. (2006). *Supply chain operational reference model, version 8.0.* Pittsburgh, PA: Supply Chain Council.

(The*) Trailer Train Story* (undated). Chicago, IL: Trailer Train Corporation.

Transportation Research Board. (2005). *Critical issues in transportation.* Washington, DC: Transportation Research Board.

Walker, M. (1994). *Comprehensive railroad Atlas of North America* (2nd. ed.). Faversham, Kent, UK: Steam Powered Publishing.

3 Securing the infrastructure

As a commuters embarking on your daily commute to work, you rarely think of the physical components of the system you are using. You arrive at the commuter rail parking lot in suburbia to board your train to the city center. Completely oblivious of your surroundings, you pass through the station platform to board the train. You settle down to read the newspaper, totally ignoring the track, culverts, bridges, and interlockings over which you are passing and the signals along the way that control the train movements. You might look out the window and see cars stopped at the railroad-highway grade crossings or notice the brief darkness as the train goes through a tunnel.

You arrive at the south terminal in the city and head to the subway to get to your office. As you get off the train, you notice Amtrak's Acela on the next track but perhaps not the overhead power lines. You proceed through the terminal area and take the escalator down to the subway, through the tunnel and onto the platform. When the train arrives, you get on and continue reading the newspaper not paying attention to the tunnel, tracks, powered third rail, and signals. You get off at your stop only two blocks away and go up the escalator and on to the street level.

If you are one of these commuters walking the several blocks to your office, you remember several things. You remember seeing freight trains along the route, combined residential and commercial development on both sides of the track, a major U.S Postal Service facility, a university, passing through several subway stations on a separate track, and crossing over a river. What you don't think about is that all you have observed is the infrastructure typical of railroad and rail transit systems. Taking this a step further, and although you don't know it, you observed the relationship among several of DHS's critical infrastructure sectors[1] and the potential for cascading impacts. If you had thought about it, you would have recognized that a natural or manmade disaster impacting the railroad could impact other critical infrastructure and vice versa.

Overview

Since you are thinking about railroad and transit infrastructure, let's look at the infrastructure and operational security, as they go hand in hand. It is important to understand that infrastructure planning and design focus on the safety and security of the constructed facility, its impact on operations, and how the passengers and workers are provided with a safe and secure environment. Infrastructure should not preclude secure operations, and operations should not compromise infrastructure safety and security, and protection.

Although infrastructure includes systems,[2] track and structures will be the focus of this chapter. We will, however, briefly discuss systems, but this subject warrants a book of its own because of, among other things, its complexity, evolving technology, and methods of compromise.

Track structure is comprised of rail, crossties, and other track material, such as rail fixation systems.[3] Also included is specialty track work, such as turnouts, crossovers, and grade crossings. Subgrade and drainage structures are also considered as complementary components of the track structure. Further, there is a link between track and signals, and train control and power, as there are components of that track structure that support the systems and vice versa.

Structures include bridges, tunnels, stations, and administrative and maintenance buildings. Security at/for structures varies. Infrastructure protection and hardening is important when considering security at bridges and tunnels. As there are different types of stations, such as terminals (i.e. New York's Penn Station), wayside along a rail passenger route, and underground and elevated typical of rail transit, the approach to security will differ, but protection will focus on the principles of Crime Prevention through Environmental Design (CPTED) (Crowe and Fennelly 2013). The principle of CPTED is to deny individuals, with hostile intent, the ability to use the structure or facility for perpetrating a crime, vandalism, or a terrorist attack.

Administrative and maintenance buildings, although not a traditional target where the intent is mass casualties, also pose a risk. Concern includes sabotaging rolling stock in a maintenance building; hence, access control and closed circuit television (CCTV) should be designed into a facility. Dispatch/operations facilities should also have access control and CCTV, as train operations are controlled and monitored there, and an intrusion could prove disastrous. This is in addition to a possible cyber attack on the supervisory controls and data acquisition system (SCADA) associated with train control. As such, system redundancy is important to the extent that operations are restored as expeditiously as possible should an event occur.

Situational awareness

Railroad track is constructed and maintained to meet FRA TRACK SAFETY STANDARDS that determine and relate track conditions to operating speeds.[4] Safety inspections are conducted at regular intervals based on FRA Track Safety Standards for visible and internal defects, deficiencies, and component deterioration. These inspections can also be used to observe tampering with the track structure or the presence of improvised explosive devices (IEDs) or other objects that could cause a derailment or incident.

For bridges and in tunnels, the structural inspection frequency is less than that for the track structure. Therefore, the track inspections can be used to observe anomalies, such as the presence of IEDs or signs of tampering with the structural components of bridges and tunnels. Then, if something is found that could indicate tampering with a structural component, a detailed inspection by a qualified inspector or engineer should be conducted. In addition to the tunnel structure, the lighting and ventilation systems should also be monitored.

At passenger stations, a similar inspection process can be used. But there will be some subtle differences depending on the station location and type, to include multimodal terminals, such as South Station in Boston or Penn Station in New York City. At wayside stations, track inspectors usually provide the "boots on the ground" observations that could identify anomalies. It is a little more difficult at underground stations, but the track inspectors can provide regular observations supplemented by the more frequent observations of the train operators. Moreover, where there are operating, maintenance, and railroad police personnel present, they can provide a second layer of observation and promote passenger awareness in the spirit of the Federal Transit Administration's (FTA) "See Something, Say Something" campaign. At multimodal terminals, where commuter rail, Amtrak, rail and bus transit, and intercity bus are present, there usually are a number of operating, maintenance, and railroad/transit police personnel capable of providing "eyes on the target". The visible presence in the station also provides a deterrent.

In any case, what is found should immediately be reported to the operations (or transportation) department and police/security via the train dispatcher or other method so that a determination can be made with regard to handling the issue. This could include diverting traffic, evacuating a station, and/or conducting a detailed inspection. In any event, the railroad/transit police or security / risk management department and appropriate maintenance department (i.e. track or bridge and building) should also be contacted and informed of the situation. If what is found is a suspected IED, other potential weapon of mass destruction (WMD), or tampering, the personnel finding the device should not attempt to move, remove, or diffuse it. If what is found is tampering with the track structure, bridge or tunnel component, or station structure component, the engineering department should be contacted to assess the condition before anything is done.

In addition to the equipment safety inspection requirements of the AAR,[5] inspections can also be used to find tampering with the rolling stock (rail cars and locomotives), presence of a suspected IED or WMD, or hidden compartments in rail cars that could carry individuals or material that could pose a threat. Inspections of the rolling stock by the train crews prior to departure and equipment maintainers can also provide additional layers of security.

Regulatory inspections are also conducted by the FRA, TSA, and Customs and Border Protection (CBP). The FRA inspections are safety focused, while TSA inspections are security related and, at this time, are primarily focused on chain of custody and the handling of TIH material.[6] CBP inspections, primarily at the border, are focused on smuggling, drugs, and human trafficking. However, each of the agencies' personnel has been instructed to report an anomaly if they see one that could compromise safety and security even though it is outside the scope of their inspection. There is a memorandum of understanding between TSA and FRA on this matter.

Shared corridors and rights-of-way

One of the principles of TSA's rail transportation security regulation addresses operations where passenger and freight trains share corridors or tracks. This is particularly relevant in areas where TIH or poison inhalation hazard (PIH) material is handled. A concern is where freight trains may be parked on a passing siding and having a passenger train pass by, creating an opportunity for a release of hazmat or an explosion affecting the passengers. The reverse scenario is also possible, where the passenger train is on the siding, and the freight train is passing by. Main tracks and passing sidings, which typically have track centers[7] of 15 feet, create a significant vulnerability.

On shared track, and because the axle loading of freight trains is greater than that for passenger trains, rail wear could be accelerated unless heavier rail is used. Hence, considering the cost of installing and maintaining heavier rail, lighter rail is often used with the result being that a defect in the track structure would be more serious for a freight train than for a passenger train. Thus, weaknesses in the track structure could easily be exploited to create an incident. This is especially relevant where the track is on an embankment or subgrade where there is construction/excavation near the tracks.

An example of the above is when, in the mid-1980s on a line shared by Amtrak and a freight railroad, an adjacent excavation near the tracks caused a subgrade failure when a freight train passed by, resulting in several cars falling into the excavation. The relevance of this is that an Amtrak train, with its lesser axle loads, passed by the excavation several hours prior without incident. What would have happened if the freight train preceded the Amtrak train and weakened the subgrade? The point to be made is that, when the lateral restraint is removed from an embankment or subgrade, an "accident" could occur. This mechanism of derailment could be mimicked by a terrorist with malicious intent, particularly in remote areas.

This concept can be expanded to include railroad bridges. In this case, a weakness on the bridge superstructure (i.e. critical member) could be exploited, to include sabotaging critical members. The substructure could also be compromised. The different speeds between freight and passenger trains operating on the same track will create different dynamic loads over the track and bridges. Bridge loading is a significant issue because forces and stresses are the basis of railroad bridge design, and it includes several components such as the weight of the bridge (dead load) and movement of the train over the bridge (live or dynamic load) (AREMA 2013). This difference could be exploited by using freight train operations and compromising critical members (i.e. via strategically placed explosives) to weaken and "take down" a structure. This is a key topic that will be discussed in greater detail later in this chapter.

Infrastructure and operational security

Railroad infrastructure should not hinder operational security. And operational security should not unduly burden the planning, design, construction, and maintenance of infrastructure. They should promote each other. CPTED and other references provide planning and design guidelines and recommendations to minimize the risk of compromising a station, structure, or facility where IEDs or other WMDs could be hidden or the structure itself could be compromised.

Train crews, railroad police and security personnel, and maintenance workers should be able to have clear sight distances to clearly observe safety hazards and places where security could be compromised. However, it should be noted – so should the passengers. Railroad and transit personnel should be familiar with the system's operations, infrastructure, and relationship between them with regard to security. This can be accomplished by formal and on-the-job training and exercises.

Regulatory process and partnered initiatives

The FRA is charged with maintaining the safe, reliable, and efficient movement of passengers and goods. The TSA, which is the transportation arm of the DHS, is charged with protecting the nation's transportation systems to ensure that the movement of passengers and goods are secure for terrorism. The FTA provides financial assistance and oversight to construct new transit (and commuter rail) systems and improve, maintain, and operate existing systems. Under its State Safety Oversight (SSO) Program,[8] the FTA also oversees the safety practices of rail transit systems in states that are eligible for funding. Although there are many other agencies that are involved in transportation safety and security, the FRA, TSA, and FTA are the primary modal agencies.

The regulatory process employed by the FRA, TSA, and FTA is operational, but there is a relationship between operations and infrastructure. The FRA, for example, uses its Track Safety Standards[9] to regulate speeds, but, by doing so, it mandates minimum track maintenance levels to safely support the speeds. TSA regulates the handling of TIH/PIH, which is operational.[10] The railroads use infrastructure, to include access control and CCTV to assist in the operational requirements of the regulation. The Safety and Security Management Plan (SSMP), which is a requirement of FTA-funded capital projects, focuses on, among other things, developing design criteria. This in turn provides for infrastructure development that promotes operational security and satisfies the management requirement of the SSMP.

Industry initiatives are a partnered approach that is exclusive of regulation and promotes a cooperative and integrated approach to transportation security. TSA has developed several industry initiatives that partner with the railroads and transit systems. They address, among other things, operations and training, but also include infrastructure components and how these factor into operational security. For example, curvature and gradient and when combined require slower speeds. When trains operate at slower speeds, they are susceptible to an incursion. And when trains are stopped at an interlocking, which is a point where rail lines/routes merge or cross requiring signals to control train movements, they are more vulnerable to incursion (Railway Age 2002). Another TSA-partnered initiative looks at adjacent land use, the commodities carried, and routing within a HTUA[11] when assessing freight rail corridors for vulnerabilities and risk.

Threat and vulnerability assessments (TVA) and security reviews

When looking at infrastructure planning design and construction, considering security is important when thinking of operational security. This "look" can be formal or informal and be taken as a responsible practitioner or as part of the grant process. Security reviews are required by DHS/TSA for projects involving critical infrastructure, but they are also a part of the SSMP process. There are many documents and references that are used to assess risk and review designs for security awareness and compatibility. The following is a discussion of a few methods and assessments used in the process.

Security focus during initial project development often begins with or is generated by a TVA. FTA, in its "purple book", addresses the threat to and vulnerability of a transit system (FTA 2001). The "purple book" is an important source document in project development that looks at the likelihood of a threat impacting the transit system or facility. It provides a five-step methodology that identifies, prioritizes, and recommends, for implementation, countermeasures to be considered during the planning and design process using sound engineering judgment, best practices, and an understanding of the system's operations. The steps are asset analysis, target and threat identification, vulnerability assessment, consequence analysis, and countermeasure recommendation.

The FTA Preliminary Hazard Analysis (PHA) is one of the tools used to assist transit systems in providing safe and secure operations (FTA 2003). The PHA was developed based on, among other things, recommendations of the National Transportation Safety Board (NTSB) and is part of the FTA grant process. The PHA is conducted during planning and preliminary design to assess the hazards associated with a project. It is intended to certify that appropriate measures are taken to ensure safety and security, an appropriate design is prepared, and procedures are developed to mitigate the hazard potential. The PHA is a seven-step process that describes events that could result in safety or security issues, assesses the frequency of occurrence, identifies that which could contribute to the exposure, describes the effects that the event, and prepares a course of action analysis that looks at worst-case scenarios and probability of occurrence. The PHA then identifies actions that prevent or mitigate the event and threat (Gordon 2014).

A formal process is the DHS/TSA Security Review for critical infrastructure and covers all modes of transportation. For new-build and reconstruction projects, the review is conducted at the approximate 30 percent design stage; and, for existing structures and facilities, it is conducted based on ranking, complexity, exposure, and particular threat. A typical security review is sponsored and conducted by TSA in cooperation with the U.S. Army Corps of Engineers Protective Design Center (USACE PDC), Federal Highway Administration (FHWA), state department of transportation, rail or transit entities, design and specialty consultants, and other parties, as appropriate, for the project. The intent of the security review is to identify design, construction and operational measures to mitigate a terrorist attack. Assault planners, which are typically former Navy SEALs, Army Special Forces, or

Force Recon Marines, identify the likely threats and vulnerabilities of the transportation facility, which the designers use to identify design solutions to mitigate the threat. For a bridge, hardening critical structural members, surveillance by way of CCTV and intrusion detection, and setback are common strategies and methods used to increase the security at and approaching the structure.

The FTA SSMP is part of the safety and security certification process for projects funded by the FTA (FTA 2007). The principles of the SSMP, however, can be used for freight rail and other transportation modes. The SSMP consists of 11 sections, which address a project's or system's overall:

- management commitment and philosophy,
- integration of safety and security into project development process,
- assignment of safety and security responsibilities,
- safety and security analysis,
- development of safety and security design criteria,
- process for ensuring qualified operations and maintenance personnel,
- safety and security verification process,
- construction safety and security,
- requirements for 49 CFR Part 659,[12] Rail Fixed Guideway Systems, and State Safety Oversight,
- FRA coordination, and
- DHS coordination.

The coordination with FRA and DHS/TSA is important because commuter rail operations in shared corridors could have freight trains carrying hazardous material stopped or passing locations where passenger trains are operating or idle. Coordination with DHS is geared to applicable TSA regulations and security directives. TSA has voluntary partnered initiatives that promote secure operations and a protected infrastructure involving rail freight hazmat operations.

Considering security during the planning and design process

Security is best introduced during project development: the planning and conceptual design phases when planning new, upgraded, or expanded transportation facilities. This is so that threats and vulnerabilities and associated risks are identified early and so that mitigation measures can be designed into a project, thus permitting funds to be programmed into the project. Operational security enhanced by physical measures is key to protecting the users of the system, workers, and adjacent landowners and to resuming operations quickly should an incident occur. There are formal processes, as discussed above, but a "seat of the pants" approach that reflects a "boots on the ground" approach of the practitioner is helpful.

When TVAs or other security assessments are conducted for a railroad, transit system, transportation facility, or metropolitan area, there is often a need to improve security after-the-fact. When doing so, the cost of retrofitting a facility is more costly than if designed at the beginning of a project. Because of infrastructure costs, security improvements for existing facilities (i.e. rail yards and stations) are often more operationally focused than infrastructure related. In this case, the railroads and transit systems are more involved because it often requires operational changes that can be costly and/or require the transportation provider to pay for a considerable portion of the necessary improvements.

In either case, close coordination with DHS/TSA, the FRA, and other involved agencies is required so that partnered initiatives are developed and an equitable cost sharing is developed. This partnered approach has worked well with DHS/TSA in that the railroads and transit systems buy into the initiative more than if regulated. Also, some of the risk is multi- and cross-modal in rail or intermodal yards and in ports and/or at border crossings. Therefore, working with the U.S. Coast Guard (USCG), CBP, etc. in developing multi-agency partnered cooperation and initiatives is effective.

It is understood that the provision to inspect 100 percent of the cargo containers in the Safe Ports Act of 2006 is not possible (Willis 2012). Therefore, the industry, to include the railroads, must explore other initiatives. One initiative is the CBP-sponsored Customs-Trade Partnership Against Terrorism (C-TPAT), which began in 2001 to protect the international supply chain. Through C-TPAT, which is one of the many cargo enforcement strategies of CBP, businesses have strengthened the security of the international supply chain (CBP 2012).[13] The freight railroads, particularly the larger ones that receive freight at the border, have voluntarily participated in this public-private partnership to ensure the security of the operation and facilitate the ease of handling rail cars crossing the border.

Track structure

Track construction practices date back to the mid-1800s. At that time, track was constructed using manual labor for the most part. Bridges, stations, and other structures were constructed using similar methods. Track is constructed at-grade, in a cut, on an embankment, on a bridge, or in a tunnel. It was originally constructed of wood crossties and "stick" rail, which is the standard 39-foot length, and other track material (OTM), such as cut spikes, tie plates, and rail anchors. The rail weight (as measured in pounds per yard) was originally less than that used today, being as low as 70 lbs.[14] and 80 lbs. per 39-foot rail lengths. This is because of the smaller rail cars and lesser weight loads in the early railroading days. Rail is currently in the 132 lb. and 156 lb. range.

In the early days of railroading, the track was constructed on a prepared subgrade and stone ballast that was interlocked for support and drainage. Culverts were installed to promote positive drainage, and bridges were installed for crossing waterways, roads, and other geographic features. As technology progressed, track construction became relatively mechanized with rail and crosstie laying machines, ballast regulators, and laser lining equipment. With the heavier axle loads and longer car lengths, rail weights have increased, concrete crossties are used in addition to wood, and OTM has stayed relatively the same with the exception of the use of screw/bolted and elastomeric rail fasteners. With the advent of technology and the heavier axle loads, track can be installed or replaced rapidly. This is important so that service can be restored relatively quickly after a derailment or incident, such as vandalism or a terrorist event.

Track construction over the past 30 to 40 years has evolved to using heavier rail, to accommodate the heavier axle loads, longer trains and larger rail cars, and the use of continuously welded rail (CWR). Using CWR requires a mechanized method of construction that is more precise than conventional track construction and reduces labor. In addition to reducing rail wear, it significantly reduces a point of failure in the track structure, namely joints every 39 feet. Not only does CWR reduce the maintenance required with joints, it takes away a repetitive point (every 39 feet) at which a terrorist could, with relative ease, damage the track structure and cause a derailment with potential disastrous results.

In addition to mainline track, there are many other track types, based on use. Mainline tracks are the major routes with high freight densities, as measured in ton-miles. The track is usually constructed of heavier rail, and track maintenance is at a higher level, as per the FRA Track Safety Standards, to support higher speeds. Mainlines are usually where freight and passenger traffic is mixed. For lesser density freight traffic, there are branch lines; where industry is served, there are industrial or spur tracks. Passing sidings or just sidings are tracks adjacent to main and branch line tracks to, among other things, place rail cars to be picked up or delivered to industry or allow one train to pass another. This configuration poses a risk when a passenger or freight train carrying hazardous material is on the siding and the other passes by on the mainline.

Yard tracks are configured as a "ladder" off a main, branch, industrial, or spur track. They are used to store railcars before delivery to an industry or after pickup and before being put in a train for transport. Yards are a major focus of TSA with regard to the handling of TIH/PIH. An individual with bad intent could gain access to a TIH/PIH rail car "buried" in the yard and open a valve to release the toxic material. To piggyback on this, the FRA also has rules regarding the length of time a hazmat car can remain in a rail yard.[15] Therefore, yards should be designed to allow personnel to easily monitor what is in the yard and deter unauthorized access or trespass.

Turnouts, crossovers, and crossing diamonds are examples of specialty track work. These components are where trains cross from one track to another or, in the case of a crossing diamond, cross another track. These components often are associated with interlockings and interface with the signal system. They require more frequent inspection and maintenance and are a point of weakness that could cause a derailment or be otherwise compromised.

Tracks at railroad-highway grade crossings involve several features that could be compromised. For vehicles to cross the tracks, timber, rubber, or concrete surfaces are placed before and between the rails. Protection to the traveling public is provided by flashing lights and gates, which are part of the grade crossing signal system. Approach circuits in the track and the signal circuitry are designed to provide sufficient time for the flashers and gates to be activated to protect the traveling public from the oncoming train. Because access to the tracks and signal casings are unrestricted, any person with bad intentions could easily tamper with the tracks, approach circuits, and signals for the crossing.

Ease of access to the tracks is a major issue that all railroads deal with. This problem is exacerbated when the track is constructed on an embankment and at grade, and when drainage structures are in the subgrade. The embankment can be cut into removing lateral support, which could cause a failure as the train passes over. The track at-grade could be tampered with, or explosives could be used to damage the track structure as a train passes over. Drainage culverts could be a place where explosives could be placed or an operative could hide, while the drainage ditch could be another place for the operative to hide. The risk is heightened at night, where it is difficult to see the tracks, subgrade, and drainage ditches. In a cut section, it is not as easy to gain access to place a bomb or other device and get out. But it would be easier to sit on top of the embankment and throw a bomb onto the tracks as a train approaches or use a standoff weapon, such as a rocket-propelled grenade (RPG).

The vulnerability of track on bridges and in tunnels, as well as the structures themselves, is somewhat less, as they are not as easily accessible. This will be discussed in the following section. And it is somewhat less of a problem with rail transit systems, as they are electrified by third rail or overhead wires and are usually fenced, restricting easy access. Commuter rail has similar vulnerabilities and risks as freight rail.

Bridges and tunnels

Bridges

When looking at the bridges along the route during your morning commute, you did not realize that early bridge construction was primarily timber and steel. Concrete and composite material bridge construction is relatively new, comparatively speaking. The major components of a railroad bridge are the superstructure, abutments and piers, and approach. Typical railroad bridge types include stone arch, truss, and deck plate girder. Abutments and piers can be stone, concrete, or steel on concrete foundations. Track is affixed to bridges by the conventional method of construction: direct rail fixation or in a ballasted section. The method of track installation is based on the type of bridge used, vertical clearance constraints, weight restrictions, and design loads.[16]

You realize that railroad bridges span rivers (the river that you just crossed), roadways (the traffic congested highway you just passed over), ravines, other railroad lines, and rail transit lines. Access to the bridge varies with the physical feature spanned and relative vulnerability present. The number of spans and total bridge length also affect the vulnerability. A bridge over a river or other waterway will have abutments and piers. The abutments are accessible from land, while the piers are usually accessible only by water. Although the abutments are more substantial than the piers, the risk of compromise is generally greater at the abutments than at the piers because of ease of access, but more explosives would be needed to damage them. Boats are generally needed to get to a pier, but fewer explosives would be needed. This is where standoff distances come into play.

For bridges over highways and rail lines, access is relatively unimpeded. For bridges spanning highways, traffic could be a deterrent. Bridges that span other rail lines will have traffic, but the deterrent is more associated with the tripping hazard the track poses, especially at night. Where a bridge crosses a ravine, access can be achieved on foot, but, based on the depth of the ravine, footing may be difficult. This would make it difficult to gain access, reducing the risk but not eliminating the vulnerability. Denying or restricting access to the bridge superstructure and substructure in any case is important in minimizing the risk of a terrorist attack. Methods of monitoring the bridges could include motion sensing, monitored and recorded CCTV, intrusion detection, and restricting access. Much of this is currently used for safety purposes.

As bridge design evolved and computers were used, factors of safety were lessened because of the preciseness of the calculations. This results in critical members[17] of the structure, which are more susceptible to damage or failure associated with tampering, such as placement of explosives, having a catastrophic failure. With the lesser factors of safety, fewer explosives may be required to cause damage or bridge failure. This, when considering standoff distances, makes the risk of an incursion greater as less explosive would be required to compromise the critical member. This concept is for steel bridges, but concrete and timber bridges pose different problems, which could be just as significant.

Tunnels

Tunnels pose different concerns than bridges. Tunnels in a rail transit environment contain underground stations that are under the city (i.e. the new PATH World Trade Center station) and cross under waterways (i.e. Washington Metro under the Potomac River) in tunnels. Underground stations are often closer together than at grade and for commuter rail. The risk, therefore, is more than just an explosion and structural damage and destruction. The threat includes the release of toxic chemicals or biological agents. The station impacts will be discussed below.

The risk of a subway train "pushing" the chemical or biological agent down the tunnel and perhaps to an adjacent station or tunnel (piston effect) is a factor that must be considered. Therefore, and in addition to monitoring, intrusion detection, and access control, ventilation is important to mitigate an event. This leads to the question of what to do about the use of ventilation to introduce a biological or chemical agent. Also, flooding the tunnel by compromising the tunnel roof is a vulnerability that must be considered. This all led to the importance of the tunnel having provisions for access by emergency responders and egress for the passengers, in the event that the station is compromised and cannot be used.

Freight and commuter rail and transit tunnels that do not have stations could and often do run below buildings and populated areas. Take Amtrak running below New York City on the Northeast Corridor or CSX freight trains operating below Washington, DC, as a prime example. The vulnerability is if the tunnel is part of or could impact the adjacent building foundations. In this instance, any explosion in the tunnel could have far-reaching impacts, to include possibly damaging or toppling the building(s) above. The ideal solution would be to isolate the tunnel from the building foundation. This is not practical unless done so during the planning and design phases of either or if it is vital to protect a building because of its use. To mitigate the risk, intrusion detection, motion sensing, monitored and recorded CCTV, and access control are traditional methods that are used deter and detect unwanted activity.

Power, communications, train control, and signals are also carried along the tracks, through tunnels, and over bridges. These systems are typically affixed to structures along the tracks, the bridge superstructure, and tunnel walls or are encased in ductwork along the tracks, or in a bridge deck or tunnel floor. Therefore, an attack on the structure could have a far-reaching impact if the systems are also compromised. Thus, when looking at the vulnerabilities and risks associated with bridges and tunnels, one must consider the potential of far-reaching impacts and the possibility of the systems being used to cause an accident in the tunnel or on the bridge.

Other structures

In addition to bridges and tunnels, a railroad or transit system will have stations, maintenance and other buildings, and signal structures within or adjacent to the right-of-way. Maintenance and other buildings include administrative buildings, dispatch or operations centers, and facilities that include locomotive and car shops. The risks associated with administrative buildings are less than for operations centers and locomotive and car maintenance shops. Why? Operations centers, if compromised, could cause "failures" in the signal or train control systems, thus causing an accident. In a maintenance facility, a bad actor could access rail cars and locomotives to either sabotage them or place an IED in or on them.

Other buildings

At operations centers where trains are controlled via the signals, any incursion could have disastrous results if the safety built into the signal system is compromised or facility is "hijacked". There are many disastrous results that could occur, one being a train running through a signal causing a head-on crash. The threats, vulnerabilities, and risks to operations via the systems will be discussed briefly in a following section.

The threats and risks associated with locomotive and car maintenance shops are related to tampering with the vehicles or placing an explosive or other device on the locomotive or rail car. Tampering with equipment could include placing an IED in a compartment of a passenger rail or transit car, sabotaging a valve on a TIH tank car, or tampering with the braking system controls on the locomotive. The most effective way of making sure the locomotive or rail car is not sabotaged is to thoroughly inspect it before releasing it into service. Deterrents to tampering with the vehicles are access control, motion sensing and monitored CCTV, and intrusion detection, particularly after hours. What about vetting the employees?

Stations

Passenger stations, like airports, pose a significant risk. The stations are either above or below ground, along the route or at terminals. Terminals, which are usually multimodal, can have surface and underground stations, such as the MBTA's North Station, which has commuter rail, a subway line, and a light rail line. For commuter rail, aboveground wayside stations are vulnerable because of unrestricted access, but ridership is relatively low when compared to terminals and downtown subway stations. This is, in part, due to the longer headways.[18] Typical security measures include unobstructed sightlines, monitored and recorded CCTV, and lighting. Security measures at subway and light rail wayside stations, like commuter rail, include unobstructed sightlines, monitored and recorded CCTV, and lighting, but have restricted access.

Underground stations, which often are primarily transit, pose the greatest risk. This is because there may be many stations in a relatively long tunnel through a city, they are spaced relatively close together, they may have more than one line in tiered levels, and, during peak periods, they run on close headways as frequently as every two minutes. This results in thousands of people in the station at one time. Underground commuter and passenger rail stations also exist, such as New York's Penn Station, where Amtrak, Long Island Railroad, New Jersey Transit, and the MTA New York City Subway operate from underground platforms.

When you are taking the escalator from the commuter rail station to the subway platform, you start thinking of the potential risks at underground stations. If an explosive device was to be employed in addition to the immediate destruction, evacuation would be difficult. There would be chaos with people evacuating, first responders accessing the station, and the hazards associated with the explosion, fire, and smoke. Structural damage to the station cannot be ruled out. There is a growing concern that a secondary attack is possible targeting the evacuees and first responders. Moreover, trains would have to be evacuated from the station and others prevented from entering and possibly carrying the fire and smoke down the tunnel and possibly to another station.

For a chemical or biological attack, the same concerns would exist as with an explosion except for the damage to the structure, fire, and smoke. In this case, there would be a concern that trains in or entering the tunnel could push the contaminant into the tunnel to "downstream" stations, exacerbating the situation. It is important that trains be prevented from entering the tunnel and that the tunnel/station be ventilated as soon as possible. Provisions must be in place to neutralize or filter the contaminant before ventilating to the surface.

In either case, ventilation, well-defined access and egress, immediate operational response so as not to send additional cars into the tunnel/station, and access control are essential. It is important that preventative measures be used to deter such an incident. This includes having the station well monitored with strategically located and monitored "in real time" CCTV, intrusion detection being used to prevent a perpetrator from entering the station through the tunnel, law enforcement visibility particularly at the points of access and egress, chemical and biological detectors, and lighting. Operator awareness and the FTA "See Something, Say Something" campaign provide another layer of security.

Signals, train control, and systems

As you sit in the subway car, you begin wondering how the trains are operated safely with such high frequency during rush hour. As stated previously, infrastructure and operational security go hand-in-hand and this infrastructure planning and design should consider both safety and security. Unlike track and structures, signals, communications, and power are more easily compromised and could result in head-on and highway grade crossing accidents, as examples of incidents. This potential is exacerbated when passenger trains operate on the same tracks as freight trains and where roadways with high vehicle traffic cross at grade. The latter is worsened where hazmat, and in particular TIH/PIH, is present on either the rail line, the roadway, or both. With the current issues with the rail transportation of Bakken Crude, the potential of an accident or targeted incident is greater. This discussion will be limited to the signals, train control, and power systems supporting freight and passenger/commuter rail and not the sophisticated train control systems for rail transit systems. That is the subject of a separate book.

As for signals, train control, communications, and power systems (hereinafter referred to as *systems*), we will look at wayside signals that control trains along the route and grade crossing signals that control trains and motor vehicles that cross each other at grade. Positive train control or PTC is a signal system that is intended to automatically stop trains before certain types of accidents occur and will also be discussed. According to the AAR, PTC, which is mandated by Congress,[19] is intended to prevent head-on and rear-end collisions; derailments caused by excessive speed; trains entering onto sections of track without authorization (i.e. where maintenance is being conducted); and the movement of a train through a misaligned or improperly aligned switch.

Signals, communications, and power are interconnected with the track and structures. Civil engineering components of the infrastructure, such as signal bridges and the physical structure of control/dispatch center, are required to support the signal, communication, and power systems. Compromising the track and structures could have a direct impact on the systems. This can be via track disruptions that could damage signal cable, destroy signal bridges, and bring down pole lines.

Considering the above, there are many locations and "opportunities" where an accident can be purposely caused by tampering with the track, structure, and systems. For example, in a remote location more than 50 miles west of Phoenix, an Amtrak train derailed. It was in the early morning of October 9, 1995, when Amtrak's Sunset Limited traveling from Miami to Los Angeles derailed in the middle of the Arizona desert. The derailment occurred as the train passed over a trestle and fell into a dry riverbed 30 feet below, killing one person and injuring 78 more.

Investigators found notes along the derailment site signed by the "Sons of Gestapo" spouting anti-government rhetoric; hence, they knew that this was not an accident. This was corroborated when, along with the notes, railroad spikes were found to have been removed from the track. From this, it was determined that whoever did this knew how to override the railroad's safety system so that neither the train crew nor the dispatcher had any idea that the track was not intact (Hajek 2015). This case is still unsolved 20 years later and is considered domestic terrorism. This raises the question of whether there is a way a signal system can detect tampering of this nature and stop the train in advance.

In the early morning of September 22, 1993, and in a dense fog, barges struck and misaligned the Big Bayou Canot railroad bridge near Mobile, AL (Smothers 1993). This knocked the unsecured end of a bridge about 3 feet out of alignment and severely kinked the track. Within minutes of the barge strike, Amtrak's Sunset Limited crossed the bridge at a high speed and derailed at the kink, striking the misaligned bridge. This caused the train to derail into the river, killing 42 passengers and 5 crewmembers and injuring 103 passengers. The NTSB determined that the probable cause of the derailment was the misalignment of the railroad bridge when it was struck. Although an accident, the mechanism of derailment could easily be replicated as a terrorist intent. How can an event of this nature be detected to prevent the disaster? Could a robust signal system have prevented the high-speed train from hitting the bridge by stopping it before?

Railroad signal basics

When discussing the basics of railroad signal systems, let's first look at the types of systems. The signal systems discussed herein are for freight and passenger rail and not transit. This is because, among other things, transit signaling and train control are much more complicated and specialized, and power is by electrification – a third rail or overheard catenary.

In simple terms and for the purpose of this book, railroad signaling can be divided into four major categories: wayside signals, grade crossing signals, detector systems, and the old, manual train order systems. Wayside signals will be discussed with regard to manual and controlled block systems, automatic block signaling (ABS), centralized traffic control (CTC) systems, automatic train control (ATC) systems and automatic train stop (ATS) systems (Solomon 2010). PTC is also discussed because it is the newly mandated system by the FRA. All of these systems provide a level of protection that controls the movement of trains.

As you pass several grade crossings on your commute, you wonder how grade crossing protection works. Grade crossing signals are comprised of approach circuits, which establish warning time, such as constant warning, predictor, and variable speed systems. Also, highway signal interconnection and preemption for highway emergency vehicles at the grade crossings are components that most people would not think of. You also think of how roadway traffic is controlled and if emergency vehicles need to cross the tracks, particularly when a train is approaching. And what if highway traffic backs up across the tracks?

Detector systems are located within the track envelope that warn the dispatcher and/or train crew of problems with the train. Typical detector systems are “hot box”, dragging equipment, and dimensional (high and wide) loads.[20] Last but not least and although not technically signals, the relatively small sections of tracks that have no signals and operate via train orders need to be considered. A small percentage of the operations nationwide and typically associated with the smaller railroads, train order operations could pose a safety and security risk. However, rail lines that are not signalized are not usually main routes and do not have a significant freight density. There is a link between track and signals, train control and power, as these are components of that track structure that support the systems and vice versa. The purpose of this brief discussion is to provide the reader with a rudimentary understanding of systems and how they are vulnerable to attack. Since signals and other systems components employ the principles SCADA, cyber attacks are real and growing threats. Cyber attacks and other methods of compromising railroad signals and systems are the subject of additional and more detailed analysis, perhaps in a separate book.

Methods of mitigating risk

As you think about risk, you ask yourself how can we eliminate it or lessen the impacts. Although discussed cursorily and in general terms previously, infrastructure protection and hardening, restricting access, intrusion detection, system redundancy, and situational awareness are the key methods of mitigating risk. Further, conducting a security review of new designs and conduct assessments of constructed facilities are other ways of mitigating risk. Then, as you are riding up the escalator from the subway station, you ask yourself, “How can risk be cost-effectively mitigated?”

Infrastructure protection and hardening

Infrastructure protection and hardening are ways of preventing a terrorist from causing death and destruction by exploiting a vulnerability based on the design and construction of track, structures, systems, or facilities. It is often accomplished as the result of a formal or informal design review for new construction or when assessing the vulnerability of an existing structure or facility to new or expanded threats. Hardening is achieved by strengthening a structure’s critical components to withstand an explosion that is considered likely based on many factors including construction, components, and material. Infrastructure protection is what deters or prevents a terrorist from gaining access to a structure or facility, and, if access is gained, it minimizes the effectiveness of the attack (i.e. limiting the number of pounds of explosive that can be carried). It includes stand-off distance, access control, and surveillance. Monitored and recorded CCTV, intrusion detection, and lighting are usual and effective methods of surveillance in addition to law enforcement.

Restricting access

By restricting access, a terrorist would be limited in what he or she could do. However, in the railroad environment, restricting access is difficult. In the transit environment, it is easier as the tracks and facilities are often fenced because of the third rail or, in some instances, overhead power. In either case, restricting access could be a combination of manmade objects and using the physical features. Regardless, people being aware of their surroundings provide an additional layer of security.

For buildings and stations, restricting access would commonly include fencing, lighting, unobstructed site lines, and access control via identification cards. To observe or detect unauthorized access motion sensing, monitored and recorded CCTV and intrusion detection are recommended. And in certain areas, the CCTV should be pan-tilt-zoom (PTZ) and be motion sensing in other less-utilized locations. Where there is a roadway adjacent to a building or driveway leading to a facility, bollards and barriers should be used to prevent getting close to the building without permission. All of this is not effective without an access control plan defining authorized and unauthorized access, what determines each, and the requirements and procedures for implementation.

Since bridges and tunnels have somewhat restricted access, portal hardening or bridge approach protection are key methods of restricting access. This includes fencing along both sides of the track approaching the structure combined with the physical characteristics of the land or waterway involved. The track structure, with its uneven surface is also a deterrent, especially at night. Sight lines approaching a bridge or tunnel should be unobstructed so as to prevent a hiding place for a device, such as an IED, or for a terrorist to wait to launch an attack. The latter can prevent tampering with a rail car, if train speed is slow,[21] or placing a device or IED on a passenger car. CCTV and intrusion detection at bridges and tunnels are also a consideration.

Restricting access to main or branch lines and passing tracks is usually a function of the physical features surrounding the track. Rail yards, however, are somewhat protected, as they have limited access due to the activity in a rail yard, closeness of the tracks, and lighting in the larger yards. Also, the yard tower often has CCTV to monitor the activity in the yard. This is especially true when the yard handles TIH, hazmat, or high-value shipments, such as tri-level auto racks carrying luxury automobiles. The best way of protecting a TIH or other hazmat car is to "bury it" in the interior tracks and monitor it via the CCTV and human surveillance. Yards can be fenced, but they are open at the "mouths". This can make it more difficult to gain access, but not restrict it. Since yard tracks are often part of a larger facility, such as a transit or railroad maintenance complex, full access control may be possible but is dependent on the size and location.

Intrusion detection

Intrusion detection is used at stations and maintenance facilities and has applications for bridges and tunnels. It is intended to detect activity in and around a facility to alert of movements that may be unauthorized. This is also intended to detect access and egress at unauthorized locations and via routes that are not typical for the facility's use. Methods of intrusion detection range from the simple, such as lighting and CCTV, to the sophisticated, which can include infrared and motion-sensing cameras and alarms.

At stations, it is used to detect someone that would be entering from other than the doors and platforms, which could include access via the track outside the station. The same applies for maintenance facilities and operations/dispatch centers. For bridges, the intrusion detection device should be placed so that unauthorized access from the track and underneath can be detected. The same applies to tunnels, but intrusion detection should also be able to detect an intrusion from above the tunnel. The effectiveness of intrusion detection in any of these cases is only when they are tied into an alarm system and/or CCTV and they are monitored. Recording could provide documentation of an incident assisting in an investigation and could be used as evidence.

System redundancy

System redundancy is a cornerstone of a railroad or transit system's *continuity of operations plan* (COOP).[22] It is the ability to maintain and restore operations in an expeditious manner so as to restore normalcy in the transit and commuter rail environment and bring goods to market in the freight environment. Both go a long way in quelling the traveling public's fears associated with a terrorist event or natural disasters.

System redundancy is a backup of operations or dispatch systems used to maintain and quickly restart railroad and transit operations. It can be used for natural disasters, system disruptions from component failures, and terrorist attacks. Even though system redundancy is technology based, locating the backup system at a different location is important. For example, the Kansas City Southern Railway has its primary dispatch system in Missouri and its backup system in Louisiana. This reduces the risk of both systems being compromised at the same time. Another solution to quickly restoring services is to have a decentralized dispatch system with the capability of one or more locations picking up for what may have been damaged or destroyed. Alternate routes should be identified and included in the COOP, but this may require the expense of maintaining the track, structures, and systems above that required for normal operations.

When considering the resumption of service under adverse conditions, the maintenance and upkeep of the rolling stock are important and not often considered. A well-planned terrorist plot could target key maintenance facilities. Without the ability to perform maintenance and make repairs at multiple locations, the recovery of operations could be slowed or not be as comprehensive as it could be. In addition to the protection of the facility, spare parts inventories should be distributed among the backup facilities.

There are no redundant bridges, tunnels, stations, or track. The redundancy is in the ability to execute a backup plan that involves, among other things, rerouting trains and transit vehicles around an incident, compensate for the loss of a station, and use bus bridges when the damage precludes an immediate execution of the same mode backup plan. Also, prioritizing and expeditiously beginning repair and reconstruction work and having the materials to do the work are important. Disaster planning documents should consider all of the above, as well as safely and expeditiously handling environmental or hazardous material contamination and cleanup.

Methods of protecting maintenance facilities and dispatch centers are similar to that for other components of a rail or transit system. Access control, however, is the primary solution. Fencing, key control, and identification/access cards are the first ring of security. Lighting, CCTV, and intrusion detection are adjunct to access control to increase the security posture of a facility. Biometrics is becoming value added in access control; the TSA and USCG Transportation Workers Identification Credential (TWIC) is a method of access control using biometrics.

Situational awareness

Situational awareness is basically being aware of one's surroundings. It can be an immediate awareness or awareness when planning and assessing a rail or transit facility, such as a station, bridge, or section of track. Operator awareness is a cornerstone of situational awareness and must be trained and exercised on a regular basis so that it is second nature. This includes pre-trip inspections and looking for things that do not look right on the train and along the right-of-way. This awareness is expanded to the passengers and railroad enthusiasts. The FTA and TSA "See Something, Say Something" campaign is successful on passenger rail and rail transit. On the freight side, there are groups of railroad enthusiasts that are used to expand the eyes and ears of the railroad in monitoring the safety and security of the rail lines. The BNSF Citizens for Rail Security, which is partnered with local law enforcement, and On Guard (for employees) programs, for example, have reported and mitigated theft of equipment, trespassers, and suspicious articles or vehicles left on tracks. It is a good adjunct to railroad and local policing and internal security programs.

Security reviews, as discussed previously, employ an awareness of the situation and potential threats and vulnerabilities in the process. By knowing what the threats and vulnerabilities are, an assessment of the risk involved can be made. Once the risk is determined, mitigation measures can be developed, which include operational measures, hardening the structure during the design process, and limiting or detecting unauthorized access. A successful mitigation strategy is a combination of measures and approaches that complement each other and not just one.

Notes

1 Presidential Policy Directive (PPD) – 21, Critical Infrastructure Security and Resilience.
2 The term "systems" is commonly used to refer to signals, train control, traction power, station and other power, and communications.

3 Fixation systems are made up of tie plates, spikes or fastening clips, rail connectors (aka joint bars), and assorted fasteners that connect rails to ties and ties to the underlying ballast.
4 Track Safety Standards, Federal Railroad Administration, 49 CFR 213.
5 Manual of Standards and Recommended Practices, Association of American Railroads, 2014.
6 Rail Transportation Security, Transportation Security Administration, 49 CFR 1580, 2008.
7 Distance between the centerline of parallel tracks.
8 49 CFR Part 659, which was replaced by 49 CFR 674 on April 15, 2016.
9 49 CFR Part 213.
10 49 CFR Part 1580.
11 As defined by DHS/FEMA Urban Area Security Initiative Program.
12 Now 49 CFR 674.
13 See Chapter 5 for a more detailed discussion of C-TPAT.
14 Rail is rated as number of pounds per yard.
15 49 CFR 174.14 requires that rail carriers forward shipments of designated hazmat "promptly" within 48 hours of receipt or origination of a shipment at any rail yard.
16 A design load is the weight a bridge is to carry, to include with the bridge itself and load of the trains themselves, to include the weight of the commodities carried.
17 A "fracture critical member" of a bridge, as defined by the Federal Highway Administration (FHWA), is a bridge component, typically in a steel truss bridge, whose failure would possibly cause a portion of or the entire bridge to collapse.
18 A headway is the interval between two (2) successive train units when passing a given point, such as a station, heading in the same direction. Urban Transit: Operations, Planning and Economics, Vukan Vuchic, 2005.
19 Rail Safety Improvement Act of 2008 (RSIA), Public Law 110–432.
20 In simple terms, these systems detect overheating of the "trucks" (rail car wheel assemblies); anything that is dragging alongside the rail cars, such as banding wire for lumber cars; and loads on the rail cars that are too high or wide for the line or a specific structure (i.e. bridge and tunnel), respectively.
21 The faster a train is moving, the more difficult it is to access it (the train), thus making it less vulnerable to intervention.
22 A continuity of operations plan (COOP), as per FEMA guidance, ensures that operations are maintained during and following emergencies.

Bibliography

American Railway Engineering and Maintenance of Way Association. (2013). *Manual for railway engineering*. Lanham, MD: American Railway Engineering and Maintenance of Way Association.

Crowe, T., and Fennelly, L. (2013). *Crime prevention through environmental design* (3rd ed.). Waltham, MA: Butterworth-Heinemann.

Federal Transit Administration. (2007). *Safety and security management guidance for major capital projects*. FTA Circular 5800.1.

Gordon, G. (2011, September 18–21). *Fixed facilities and infrastructure protection and passenger rail and rail transit security*. Proceedings of the American Railway Engineering and Maintenance of Way Annual Conference, Minneapolis, MN.

Gordon, G. (2014, April 7–9). *Introducing security, infrastructure protection and hardening, and system redundancy into transportation facilities projects*. Colorado Springs, CO: The Infrastructure Security Partnership, Society of American Military Engineers.

Hajek, D. (2015, April 12.). 20 years later, Sabotage of Amtrak's sunset limited still a mystery. *All Things Considered*, National Public Radio.
Railway Age. (2002). *Comprehensive railroad dictionary*. Omaha, NB: Simmons-Boardman.
Solomon, B. (2010). *Railroad signaling*. Minneapolis, MN: Voyageur Press.
Smothers, R. (1993, September 23). Dozens are killed in wreck of train in Alabama Bayou. *The New York Times*.
U.S. Customs and Border Protection. (2012). *C-TPAT program benefits: Reference guide, U.S.* Washington, DC: Customs and Border Protection.
Willis, H. (2012). *Revisiting 100% inspection*. Santa Monica, CA: Rand Corporation.

4 Railroad policing efforts

From the very beginning of rail transportation, criminal elements in society have sought to prey on rail freight and passengers, the baggage and personal effects that those passengers may have in their possession, and the physical plant, whether it be the locomotives and rolling stock or the supporting infrastructure comprised of rights-of-way, tracks, bridges and tunnels, communications and signaling systems, and, in more recent times, information technology (www.therailroadpolice.com).

With the early decisions to employ various personnel to defend against these elements came the understanding of how policing needed to emerge and develop as the U.S. rail network expanded. Railroads were seen by criminals as easy targets given that they must operate on fixed rights-of-way and often on published schedules, making for predictable targets. Moreover, a train, unless empty, represented a mass of wealth that was often traversing isolated territory. As Table 4.1 readily reveals, the need to provide a protective presence rapidly grew, keeping pace with the extent of the network, and not only changed over time but became considerably more encompassing as more and different threats to railroads became evident.

In today's environment, with terrorist attacks against all forms of transportation and physical infrastructure possible or even likely, the role of railroad policing has been transformed. While traditional roles of railroad police remain – largely surveillance of the physical plant and railroad equipment and guarding against individuals trespassing on railroad property – a new role is emerging: protecting against the possibility of terrorist attacks, and responding to threats and actual catastrophic events. To counter terrorism, railroad police units have become the railroad industry's links to intelligence agencies of government, and state and local police and homeland security units through fusion centers and planning activities. This new role has forced the railroad industry, both freight and passenger operators, to rethink the role of railroad police and attract and train individuals capable of exercising the new responsibilities in the New Normal of post-9/11 anti-terrorist policing.

The term *railroad police*, at least within the context of this book, includes the terms *special agent*, which is the preferred designation among the freight railroads and Amtrak, and *transit police*, usually making reference to those working for subway lines, light rail systems, and commuter lines.

Table 4.1 Railroad development and policing[a, b]

Decade	*Track miles*[a, c]	*Stage of railroad development*	*Impetus of railroad policing needs*
1820	<100	Baltimore & Ohio RR incorporated in 1827	
1830	1,000	Baltimore & Ohio RR built	Protect passengers, their baggage, and freight
1850	9,000	U.S. westward expansion	Labor unrested is added to the list of policing issues
1860	30,000	Railroads are a key asset to the Union's ability to prosecute the Civil War	Right-of-way security becomes a key interest
1900	130,000	Principal U.S. mode for long-distance transportation for both freight and passengers	WWI saw a dramatic increase in rail volume; railroads nationalized under USRA
1920	260,000	High point of number of railroad track miles	Petty crimes at depots and stations are a concern
1940	234,000	Heyday of the railroads; zenith of passenger miles traveled and ton-miles hauled	WWII saw another dramatic increase in volumes of freight and passengers; U.S. Army assigned to protect several key infrastructure locations
1960	218,000	Railroads appear to enter an era of severe decline; bankruptcy of lines commonplace	Advent of intermodal operations and theft of high valued goods moving as TOFC and later as COFC
1980	185,000	Economic deregulation and advent of shortline growth; era marked by reduction in track miles and number of employees	Vandalism becomes more widespread with defacing of railroad property as well as theft of freight and railroad property (e.g. signaling wire, signage, and many items seen as valuable antiques)
2000	171,000	Significant consolidation among Class I railroads as well as aggregation of shortlines into groups and holding companies, the most significant being RailAmerica and Genesee & Wyoming Corporation	Protecting infrastructure from terrorist threats, but also the infrastructure is now more broadly defined to include the underlying information systems

Data sources and notes: [a]Lieb, *Transportation*; [b]www.therailroadpolice.com; [c]Association of American Railroads

This chapter is divided into four distinct sections. First, the development of the railroads in the U.S. is discussed, focusing on the evolution of the needs of railroads for security services. Second is an overview of how railroad policing has evolved to meet those needs over the course of over 150 years of railroad development in the U.S. With the history and development of railroad policing established, the third section will discuss the contemporary issues of securing railroads in today's environment, most notably the advent of their being likely targets for terrorist activities worldwide. Finally, we present an overview of the challenges that this changing emphasis poses for the railroad police, with some examples of how major rail operators have adapted to the new situations posed by the post-9/11 era.

History

Many of the earliest trains in the 1830s and 1840s were meant exclusively for passenger travel, equipped with coaches that were little more than stagecoaches fitted with flanged wheels for use on the rails. Initial speeds were slow. As the technology progressed rapidly, locomotives became more powerful, more reliable, and faster; track structures simultaneously improved, providing safer travel at higher speeds; and railcars became heavier and more varied (Vance 1995). By the mid-1800s the railroad had quickly evolved into the transportation mode offering the greatest capacity to handle large volumes of passengers and freight, the most capable mode to handle freight of a wide range of sizes and types, and the fastest mode, a distinction that it would hold for nearly a century (Cunningham 1951).

A watershed year was 1838, when Congress designated the railroads as the primary carrier of the U.S. mail. Thus, early on, trains became appealing targets for thieves, and mail clerks were armed and deputized to guard their precious cargo. As rail transportation improved, it became the mode of choice for a wide range of freight that included not only low-value and large-volume commodities such as coal but also high-value goods of every description, including gold for the developing banking system.[1] As the railroads developed, so did the size and scope of their policing capabilities, as shown in Table 4.1.

At first, the railroads sought to protect their passengers and the luggage that accompanied them, but soon they acknowledged a need also to protect the freight that was increasingly moved in their care.[2] The first obvious solution was to arm railroad personnel whether they were engineers, conductors, trainmen, station agents, or maintenance-of-way workers. With little to no training, this approach proved marginal at best, especially as the criminal element became more aggressive as well as inventive in its methods. As labor unrest became more commonplace and spread across firms in the basic industries of steelmaking, coal mining, and railroads, it was the railroad management that employed professional security personnel who were not likely to be members of the transportation trades (www.therailroadpolice.com).

The protection of passengers from major crimes continued to plague the railroads as the volume of riders, whether long-haul passengers or commuters, continued to increase, reaching its high mark in the 1940s. The challenge was to deter unwanted elements from frequenting stations, key terminals, and other operational locations. The economic crisis after 1929 exacerbated problems of trespassing and "riding the rails" by unemployed workers or youths seeking to use trains as a means of traveling to seasonal work. Preventing panhandling, vagrancy, and pickpockets became a daily mission for the railroad police, added to the continued responsibility for protecting passengers, railroad employees, and company property (Donzalski 2009).

After World War II, changing demographics and urban decay made vandalism of railroad property – whether of cars, locomotives, stations, buildings, and other infrastructure – become commonplace. Yards and even passenger stations were often located in once-safe areas that had become blighted and crime-ridden. Some train crews knew of locations where they might be assaulted with rocks thrown from bridges or shots fired at trains. Some railroads even fitted locomotive windscreens and caboose windows with heavy metal bars to protect train crews. Sometimes vandals would damage railroad property, such as throwing switches or placing heavy objects on the rails to cause derailments that threatened the loss of life. Most often, however, the vandalism was in the form of graffiti, especially after the introduction of inexpensive spray paint cans. The results were not only unsightly but often a symptom of other forms of criminal or gang activity. In the 1960s, more and more railroads, especially in the Northeast, were declaring bankruptcy, cutting back on services, deferring maintenance of equipment, and abandoning property. Vandalism in all of its forms seemed to further contribute to a public perception that railroading was a transportation mode that was obsolete and in a state of continuing decline and decay.

While many aspects of railroading were in decline in the 1960s and 1970s, a new source of traffic, intermodal freight, began to transform the industry. The railroads had experimented with intermodal operations even as early as the mid-1800s, but fears of railroads using highway trailers to monopolize surface transportation (and to an extent bus and air passenger business) led to laws and regulations largely preventing railroads from running or participating in trucking. This began to change following World War II (Cunningham 1951). What occurred was a happy marriage of new technology and a move by government from close regulation to more market-based policies enabling the intermodal business to expand.

Several approaches to intermodal freight movements emerged. One, the so-called "piggyback car", was a flat car with attachments for regular highway trailers to be loaded, either singly or in pairs. In the 1960s, Malcolm McLean[3] began with the design for a highway trailer that was split into two components: an aluminum box and a detachable wheeled chassis. The boxes could be loaded aboard ship for efficient movement of cargo by sea and easily transferred from rail to highway movement. Advances were made in loading and unloading facilities and equipment to save time and labor in moving boxes from one mode to another. This new technology helped launch the firm Sea-Land Service, and it was emulated by numerous other ocean carriers to become the intermodal standard. After passage of the Staggers Act in 1980, the railroads were increasingly looked upon as the preferred source for long-distance land-based movement, and ports and rail facilities were refined for this purpose.[4] Intermodalism was a significant development for policing. Intermodal facilities were potential targets, given the known concentration of the high-value freight present. Railroad police had to be concerned with protecting not only trains, stations and terminals, other infrastructure, passengers, and employees but now freight that was loaded into vehicles owned by firms other than the railroads. Routine checking of bills of lading and other aspects of intermodal cargo monitoring became one of the most significant activities for railroad police by the 1980s and beyond.

The first decade of the new millennium was a time of transformation for railroad policing as rail systems outside the U.S. became targets for terrorist activities and concerns were raised that any mode of transportation was a likely target for terrorist attacks. From a freight standpoint, the U.S. rail network had been a potential target during World War II for enemy agents to disrupt the economy. But the terrorist attacks in recent years were significantly focused almost entirely on passenger trains, especially crowded terminals and commuter trains operating during peak rush hours in major urban centers. Protecting passengers took on a different perspective. It was no longer threats to passengers because of the valuables in their luggage or on their persons, but rather the passengers were the targets in their own right. In a proactive sense, freight was a similar target if it had some destructive potential such as hazardous chemicals – commodities long favored for rail movement because they were off the highways. Hazmat – those flammable, poisonous, corrosive, oxidizing, or even explosive materials – are of interest because of the threats that they pose to the safety and wellbeing of the public at large. Railroad police had to begin to think more proactively and to prioritize which sorts of threats were most likely to occur and what their consequences would be for the railroad and the communities in which they operate.

Evolution of railroad policing

Although railroad police are deputized peace officers, policing railroads has little in common with community policing for a municipality or even a state. First, when the concept of railroad police was initially envisioned, all railroads were private sector firms, usually set up as for-profit public corporations. The rights-of-way, their tracks, as well as their other infrastructure were on private property. In this sense they were similar to other types of industries that were privately owned and operated, but their geographic scope and diversity of operations and equipment meant that operations might be spread over hundreds of miles in two or more states (Donzalski 2009).

The status of the corporation is of major consequence in light of the geographic limitations that private property imposes on the ability to carry out effective policing. While the original role was similar to the modern-day security guard or night watchman, individual railroads sought to expand the authority of their police forces by reaching out to local public law enforcement and having their police deputized. This was effective perhaps initially, but it became extremely burdensome as railroads expanded, merged, and covered a large number of jurisdictions. For example, one needs only to consider the number of counties and municipalities that one of the earliest railroads, the Baltimore & Ohio, encountered on its route between Baltimore and Chicago. The B&O set up its police department in 1850 and soon recognized the complexity of working with online jurisdictions. It was not just the fact that there were many different jurisdictions, but that each needed to establish its own agreement and, potentially, limitations to be placed on its police officers. Table 4.2 summarizes the evolution of railroad policing activities.

It was not until 1865, when the Pennsylvania legislature enacted the Railroad Police Act, that railroad police were provided with state commissions issued by the governor that bestowed police powers in every county where the railroad runs, or more importantly even *where it intends to run.* The Pennsylvania statute became the example upon which many other states modeled their laws governing railroad police, a fact that simplified the pursuit of lawbreakers crossing not only city and county boundaries but state lines as well. The law was updated in 1913, but the general theme was the same: railroad police officers have broad powers, including that of arrest within their respective states of operation (Donzalski 2009).

As previously mentioned, most railroads are private corporations, or they were until 1970. When Congress relieved the railroads of their long-distance passenger business with the creation of the National Rail Passenger Corporation, better known as Amtrak, through the passage of the Rail Passenger Service Act, it established the right for Amtrak to establish a police force. Six years later Congress merged the six bankrupt freight railroads of the Northeast – Penn Central, Erie Lackawanna, Reading, Central of New Jersey, Lehigh Valley, and the Lehigh and Hudson River – into the Consolidated Rail Corporation, or Conrail. While not having long-distance passenger operations, Conrail at the outset had extensive commuter passenger operations in the New York City metropolitan area, New Jersey, and greater Philadelphia. Congress severed these from Conrail in the early 1980s, with the result being the formation of the Metro-North Commuter Railroad (MNCR), Jersey Transit Rail Operations (NJ Transit-Rail), and the Southeast Pennsylvania transportation Authority (SEPTA). Each then established its own railroad police agency (Saunders 1978).

By the 1950s, states had placed expectations on the railroads to improve the level of police professionalism. Officers were required to complete specified courses of study at the same police training academies that the public law enforcement personnel attended. Some railroads, as was the case with Conrail, went one step further and created their own academies. Some acquired such good reputations that they were frequently contracted by local and county governments to train their officers (Donzalski 2009).

In the late 1990s, Congress finally established a national standard for states to recognize the role that railroad police play. The Uniform Crime Control Act of 1990 took essentially what had been established by the Commonwealth of Pennsylvania in 1865, updated it, and applied it nationwide. Part of the rationale for this action was the continuing merging of the railroads, especially through

Table 4.2 Railroad policing initiatives

Year(s)	*Policing initiatives*	*Enabling legislation and/or developments*
1835	Informal policing by operating employees	
1850	First railroad police department established by Baltimore & Ohio RR	
1865	Contracted police services provided by outside agencies, the most famous being Pinkerton; focus was on solving robberies from railroads; watchmen were hired to protect property	Pennsylvania enacts Railroad Police Act commissioning railroad police officers; PA law becomes the model replicated by several other states as well as provinces in Canada
1870–1902	Special units established (e.g. Union Pacific's Bandit Hunters and Missouri Pacific's Secret Police)	
1915–1918	Railroads taken over by USRA during WW I	
1920	U.S. Marines assigned to trains carrying the mail	
1930	Vagrancy and trespass grew as a consequence of the Great Depression	
1940–1945	Infrastructure protection during WW II included Army detachments at critical locations	
1950–1970	Railroad police agencies became more professional; training mostly equivalent to municipal officers	Rail Passenger Service Act establishes Amtrak's authority to have railroad police; K-9 units became ubiquitous
1980–1990	Railroad police departments now highly professional with some Class I carriers developing their own academies	Federal Omnibus Crime Control Act passed providing uniform recognition of railroad police jurisdiction
2001+	September 11 changes the nature of infrastructure protection	Transportation Security Administration established within the Department of Transportation and later moved to the Department of Homeland Security; Amtrak establishes an Intelligence Unit; CSX and BNSF create their own SWAT teams; BNSF, Amtrak; and Norfolk Southern develop programs for citizens to participate in the security effort

Data source: Plant and Young 2007

the mergers of the large Class I carriers. The thought at the time was that with only seven remaining Class I railroads, the number of state jurisdictions covered by each firm presented a level of complexity to railroad police operations that required a nationwide approach.

The events of 9/11 not only changed the character of air travel, as many travelers encounter on a frequent basis, but changed transportation in general. Although the TSA has a ubiquitous presence at the nation's airports, it was also given the mission of protecting the nation's railroads, ostensibly for freight as well as passengers. This new era is one that continues to evolve.

Evolving challenges

Looking back at history, although the challenges to railroad security began at a modest pace, they quickly escalated in the lawless era as the nation expanded westward after the Civil War. On the one hand, the railroads were on the leading edge of the country's development, but that otherwise enviable position meant that the railroads were, in many instances, the only law enforcement agency to be found. This is in stark contrast to the more urban parts of the country where railroad police were required, prior to the 1865 Pennsylvania statute as well as those of other states, to secure the cooperation of the local police for taking possession of arrested suspects (www.therailroadpolice.com). Table 4.3 summarizes how these challenges for protecting railroads have evolved during the past 180 years.

Much of the earliest significant railroad police activity was actually outsourced and, in fact, helped make the reputations of such infamous lawmen as Allen Pinkerton. Pinkerton, whose fame in the security business began while solving major robberies from the railroads, established the idea that rail security was vital to the national interest (Donzalski 2009). Other railroads, the Union Pacific in

Table 4.3 Principal challenges facing railroad police

Era	*Enforcement focus*	*Challenges*
1830–1850	Theft of freight and passenger baggage	Lack of personnel, inadequate training
1850–1875	Robbery and murder of passengers and employees	Geographic coverage, hazardous duty, lack of coordination with local law enforcement
1875–1920	Theft of freight, passenger baggage, and mail	Geographic coverage, interstate jurisdiction
1920–1960	Vagrancy, trespassing, petty crimes, theft of freight	Diminishing resources with beginning of economic challenges to the rail industry
1960–1980	Theft of freight and railroad property	Too few officers for scope of network, lack of coordination with local law enforcement
1980–2000	Theft of freight and railroad property, trespassing	Too few officers for scope of network, lack of coordination with local law enforcement, expanded role for safety
2000+	Potential for terrorist attack, theft of freight and railroad property, trespassing	Too few officers for scope of network, lack of coordination with local law enforcement, expanded role for safety, too few intelligence assets given the need to become proactive in the face of global terrorism

Data source: www.therailroadpolice.com

particular, developed what in modern times might be considered a SWAT team, a group of marksmen who were also expert horsemen and assigned to a special train that was given priority for reaching any part of the company's network in the event of an incident. It was a such a team of relentless pursuers that was immortalized in the movie *Butch Cassidy and the Sundance Kid* – "Who are those guys?" Indeed, mention is seldom made that many of the famous lawmen chronicled on television, such as Bat Masterson and Wyatt Earp, actually spent significant parts of their careers in the employment of the major western railroads, including the Union Pacific and the Great Northern (www.therailroadpolice.com).

Previous mention was made of the U.S. Marines guarding mail trains immediately following World War I and again during the late 1920s. Marines were never under the command of any railroad and would be assigned only to those trains actually having United States mail onboard (Campbell 2001). The other assignment of the U.S. military was in the permanent basing of a company of infantry at the Pennsylvania Railroad's Horseshoe Curve near Altoona, during World War II. As discussed earlier, although the saboteurs never perpetrated any damage to the railroads, the FBI did capture several potential saboteurs who had been put ashore by the German Navy with orders to destroy major railroad infrastructure such as the Horseshoe Curve on the Pennsylvania Railroad and the huge Limeville Bridge over the Ohio River of the Chesapeake & Ohio Railroad. Interestingly, their goal was to disrupt the U.S. economy and war effort by attacking freight rail activity, rather than terrorizing the citizenry by attacks on passenger trains or stations.

During the era when railroad activity was at its zenith, the railroads suppressed vagrancy and petty criminal activity by increasing their police presence. As a result, railroad police employment reached its peak just after World War II. As railroad fortunes dwindled during the 1950s and 1960s, the numbers of railroad police did as well – their employment exceeded 9,000 in 1940 but has shrunk to approximately 2,500 today. With 170,000 miles of track; 160 intermodal terminals; nearly 100,000 bridges, tunnels, and culverts; and myriad passenger stations representing all sizes and descriptions, the idea of being able to police every item of infrastructure is not practical. The landmass of property owned by the railroads constitutes an area equivalent to that of the State of Delaware. It is in dense urban areas and the most desolate and unpopulated areas of the country. For this reason policing activities tend to focus on the critical nodes, stations, terminals, bridges, and tunnels, rather than on the links or lines themselves (Plant and Young 2007).

As railroad police departments have become ever more professional, they have acquired many of the tools and tactics of their public sector counterparts. They employ such innovations as K-9 units, night-vision equipment, better personal weapons, have begun using drones for surveillance and improved means of communications. Up until the past ten years, however, most of the focus remained on protecting freight from theft, passengers from a range of criminal acts, and railroad property from trespassers. This last point cannot be overstated. Trespassers may not only seek to steal or otherwise vandalize railroad property, but too often they become injured and file costly lawsuits as a result. After 9/11/01, trespassing has taken on a more sinister possibility: checking out soft targets for possible terrorist attacks.

While it is often thought that the events of 9/11 forever changed the character of air travel, every other mode of transportation was seen in a new light as well. Previously, railroads were considered strategic assets during wartime or conflict that enabled the country to move materiel and troops for the prosecution of war; hence any disruptions of that ability were seen as a strategic move that limited an adversary's ability to do so. This realization dates back to the U.S. Civil War. Ironically, the Confederacy made the first use of railroads for military success in the first battles, moving men quickly from one region of Virginia to another to tip the balance. But it was the Union that possessed the overwhelming amount of track miles and modern rail equipment, which over the course of the war enabled it to move its military assets much more quickly than the Confederacy and hence became the target for rebel saboteurs (Vance 1995). In all wars since then, railroads have played a key role in national defense and actual warfare.

While the railroads remain critical to the functioning of the contemporary economy, a new use of railroads in conflict has emerged. The *New Normal* of today is characterized by asymmetrical conflict, most often involving non-state organizations and actors, who aim to spread terror to achieve their goals. Terrorists have begun to see the railroads as ready targets where death and injury to passengers and bystanders can be easily achieved. Such attacks, summarized in Table 4.4, make for great headlines that serve to further underscore the message

Table 4.4 Recent attacks on railroad infrastructure

Year	*Event/railroad/location*	*Perpetrator*	*Impact*
1942	Targeted infrastructure destruction at Horseshoe Curve, PA	German Nazi agents	Unsuccessful; agents apprehended
1977	Hijack and passenger hostage taking, Nederlande Spoorwegen, The Netherlands	South Maluccan separatists	Train assaulted by Royal Dutch Marines
1995	Poisoning of Tokyo subway passengers	Aum Shinrugo	Sarin gas attacks
2004	Madrid commuter train bombings, Spain	Al Qaeda	Coordinated attack on multiple rush hour trains
2004	Moscow Metro bombings (February)	Chechen agents	In-station bombing
2004	Moscow Metro bombings (August)	Chechen agents	In-station bombing
2005	London Underground bombing	Al Qaeda	Multiple station bombings
2009	Russian passenger train derailment between Moscow and St. Petersburg	Chechen agents	Bomb placed on rails to cause derailment
2010	Moscow Metro bombings (March)	Chechen agents	In-station bombing
2010	New York City bomb plot	Al Qaeda	Plot foiled
2010	Russian freight train derailment (April)	Islamic terrorists	First freight train attacked in recent history

Source: Plant and Young 2007

of the attackers. By disrupting normal life, they also create great damage to the functioning of the economy and to the political life of nations. This new dimension of threat does not displace the earlier ones, but merely adds to the challenge.

Historically, the railroads were also concerned about threats from within, most notably the theft of freight or passenger baggage by employees, but the twentieth century has also brought about a concern for another threat that can be either internal or external, namely the cyber attack. Modern railroads depend on information technology to handle the myriad transactions that involve passenger ticketing, initiating and terminating freight moves, the exchange of railcars with other railroads, billing and accounts receivable, payroll, the ordering of materials and supplies, and the payment of suppliers. However, they are equally dependent on information systems to control the movement of trains across their networks and to locate locomotives and individual railcars in real time. Threats to information technology capabilities are routinely tracked by security units within IT departments, but when criminal activities are suspected, specialized railroad police units are included in the investigations.

Contemporary policing issues

The approaches taken by both freight and passenger railroads are varied but can be divided between pre-event intelligence gathering and post-event incident response. In the beginning, railroad policing involved responding to crimes already committed and only later evolved into a prevention mode that was substantially supported by visible police presence. With the threat of terrorism has come the need to gather information about the potential threats. Three different and distinct alternatives have emerged, as discussed below: two by the initiative of individual railroads and one as a public-private partnership.

Burlington Northern Santa Fe's Citizens for Rail Security (CRS)

For a variety of hobbyists, railroads have long been an attraction, with a range of seemingly diverse interests that include photography, model railroads, recording car and locomotive sightings, collecting railroad-related artifacts such as locomotive bells, station signs, and timetables, as well as interesting the casual observer through museums and historic railroads devoted to preserving railroad history. The casual observers are not likely to possess a detailed understanding of railroad operations, locomotive and rolling stock types, track structures, and train speed (to name but a few), but serious railfans, the moniker by which they are known, often do. Many railroads have viewed these railfans as irritants and potential trespassers, and, indeed, some do fall into that category with some being arrested each year as a result of trespassing on railroad property. Following 9/11, Amtrak and some of the commuter railroads established policies forbidding railfans from photographing trains and infrastructure while on railroad property and have even harassed those who attempted to do so while on public property or as passengers holding valid tickets. Some freight railroads also took action to discourage photography, and many stepped up police activity to thwart trespassing.

As years passed with no indication that such activities led to terrorism, a more relaxed approach to photography outside the actual property line of the railroad is the norm, especially since it is an established right to take photographs from public property.

The BNSF Railroad, operating across wide expanses of open territory in the western U.S., took a different position that recognized the inherent knowledge that these railfans possess. BNSF established a unique version of the *neighborhood watch* that has proved valuable in numerous communities nationwide. The Citizens for Rail Security program was established, whereby railfans can register on a BNSF website, where they not only provide personal identification and contact information but also agree to abide by certain regulations concerning trespassing on BNSF property and contacting BNSF police in the event that they observe anything that appears unusual or out of place. To date, the railroad has been pleased with the outcome, citing safety and in a few instances criminal activities that would have otherwise gone unreported. Moreover, with several thousand persons registered in the program, there is a noticeable improvement in the behavior of the railfans. As a result of its success, both Amtrak and Norfolk Southern, after at first scoffing at the idea, have launched similar efforts.

Amtrak's Office of Strategic Security and Special Operations (OSSO)

Not long after 9/11, Amtrak management realized that it needed to be more proactive with regard to the potential for terrorist threats. The result has been the establishment of a special office not only to collect intelligence but to be prepared to deal with the most severe threats such as attacks on trains and even hostage taking. The intelligence-gathering aspect of this new organization included the recruitment of individuals with specific backgrounds to deal with data collection and analysis. Many of those hired have had prior employment with such agencies as the Central Intelligence Agency, Defense Intelligence Agency, and the Office of Naval Investigations, to name just a few.

A special weapons and tactics unit was created at the same time, but, because of the specialized skillsets required, it reported to OSSO and outside of the Amtrak Police Department. This organizational structure became a source of conflict with Amtrak police, in particular with its local chapter of the Fraternal Order of Police, its bargaining unit. During the summer of 2009, after what is believed to be pressure from the Obama administration, management was forced to abandon the intelligence unit and transfer the SWAT team.

Special weapons and tactics

Despite the freight railroads keeping a relatively low profile on the subject, the SWAT team has become a fixture of BNSF, CSX, and Union Pacific. With paramilitary training much like their public police counterparts, these units have been employed in close coordination at major public events that include BNSF's unit participation at the 2008 Democratic National Convention in Denver and the use of the CSX team at the G-20 Summit in Pittsburgh in 2008 and the Obama inauguration in 2009. Union Pacific Railroad maintains a Special Operations Response Team, which, despite the different terminology, has a similar mission and training as a quick response team (www.therailroadpolice.com).

Perhaps the most unique effort is that of the CSX Team, which maintains its own dedicated train used for training throughout its system as well as for use in responding to major incidents. The CSX train almost seems a throwback to the dedicated special train created for deploying Union Pacific's Bandit Hunters during the nineteenth century. Nevertheless, the CSX effort has been used primarily for training state and local police as well as other first responders in how to deal with a range of railroad-related situations (Elliott and Flake 2010).

Fusion centers

Ever since the establishment of the first railroad police, the challenge has always been with coordinating the efforts of the private sector with its public sector counterparts. The early Pennsylvania statute resolved some of the jurisdictional issues, but even during the years after the 9/11 attacks the coordination of railroad police with other law enforcement has been a spotty affair that was heavily dependent upon local practice and a myriad of ad hoc, dyadic relationships.

After 9/11, Congress forged the Department of Homeland Security from an array of agencies that included Customs from Treasury, Border Patrol from Justice, the Federal Emergency Management Agency that reported directly to the president, and the Coast Guard and TSA from Transportation. The objective was to co-locate those otherwise disparate agencies into a single organization that could focus on defending the U.S. from internal threats. Even with a single federal agency, TSA, designated to play the leadership role, there was the matter of sharing information with agencies at the state and local levels. Moreover, there was the issue of coordination with key private sector parties that had ownership of major elements of the economy, such as transportation and key industries with security interests such as chemical manufacturing (GAO). Using modern information-sharing and communications technologies, the result is what has become known as the *fusion center*. Organized primarily on a regional or statewide basis, fusion centers share information among members drawn from a wide array of public and private entities, including the railroads operating in that region or state. Threat information and incidents occurring in one locale can be shared with others with the hopes that the inability to "connect the dots", as was the case with 9/11, can be avoided in the future (Modafferi and Bouche 2005).

A task too large

The key question concerning the ability to police the nation's railroads for threats both large and small in the twenty-first century is quite simply a case of: "Is the

Table 4.5 Protection priorities of freight and passenger railroad police

Issue	*Freight railroads*	*Passenger railroads*
Terrorist attack	• Infrastructure • Information technology	• Infrastructure • Onboard trains • Information technology
Protecting passengers		• At terminals and stations • Onboard trains
Protecting freight	• Theft protection with an emphasis on intermodal	• Theft prevention of passenger property including baggage
Trespassing	• Rail yards and intermodal facilities • Offices • Rights-of-way	• Terminals and stations • Rail yards • Rights-of-way • Offices
Vagrancy and petty crimes	• Rail yards and intermodal facilities	• Terminals and stations • Rail yards

Data source: Improving America's Security Act

network too diverse with regard to mission (freight and passenger transportation), geographic location (coast-to-coast and northern border to southern border), and cutting across so many political jurisdictions to be effectively protected?" Part of the answer lies with the professionalism, training, and tools that the railroad police have amassed, but part may also lie with the ability to coordinate efforts across organizations with the railroad industry, across other industries, across law enforcement agencies, and across intelligence agencies whereby a comprehensive picture of potential threats can be presented (GAO). Elements of each of these can be found with organizations such as the Association of American Railroads and the American Shortline and Regional Railroad Association. It can be found with public policy initiatives that seek to weave the capabilities of the private sector, in this case the railroads, into the fabric of a cooperative endeavor with public sector law enforcement. Table 4.5 outlines the shift in railroad policing priorities.

Effective protection of the railroads cannot be achieved without the participation of those that the railroads serve, which by definition must include the traveling public as well as the shippers and consignees with interests in the freight being carried. Moreover, if railroads are such a key element with the greater U.S. economy, then engaging anyone with eyes and ears as sources of intelligence cannot be an opportunity too readily dismissed. The effort of BNSF's CRS project as well as those other railroads that have chosen to follow that lead represent a set of resources that others may yet wish to examine and subsequently add to their own security frameworks even if it does represent a shift in the industry's traditional paradigm (www.bnsf.com).

Summary

In the U.S., railroad police activities are nearly as old as the railroad industry itself. Begun as an equivalent of a watchman function, policing evolved into full-fledged professional police agencies to solve crimes against railroads and to deter them wherever their railroads' businesses found them. As shown in Table 4.5, the protection priorities have expanded in number and significance in parallel with the overall role of the railroad as a key mode of transport. This evolution is now approaching a period spanning two centuries. The skillsets needed to provide for the protection of the railroads were initially outsourced, but, in today's more complicated milieu, in-house, specialized, and highly trained railroad police agencies are required.

As the railroads became more complex with respect to geographic coverage, increased speed of service, and became more capable with improved equipment, the role of the railroad police also became more complex. Protection of passengers and freight were a top priority with much of the police resources focused on those nodes where these were clustered. In wartime situations from the Civil War through World War I and World War II, protection of railroad property focused on physical assets including locomotives, rolling stock, and infrastructure. With the advent of the undeclared wars and internal conflicts, where attacks on the railroads became the basis for achieving publicity for a group's cause, protection of railroad property took on a new meaning. No longer was the railroad itself the only target; now it was also the public riding trains and gathering in stations and on platforms as well as adjacent property. The capability to anticipate possible threats and react quickly and positively to catastrophic events is now the mission of railroad police, working with partners in local, state, and national government agencies and their colleagues in other railroad police agencies.

As current threats continue to unfold, the need to protect the nation's railroads will not diminish, and the ability to do so will move beyond the actions of any individual railroad. Public-private partnerships between industry and government as well as between different levels of government and other industries will increasingly become the norm. The role of the fusion centers as information collecting and sharing assets will continue to evolve as advances in information technology and communications are brought further to bear on the problems of protecting the nation's critical railroad infrastructure.

Notes

1 Note that the U.S. was still on the gold standard at this time, meaning that the currency was, in fact, backed by the precious metal.

2 The U.S. Post Office Department founded the Postal Inspection Service (IS) in the 1830s, and this resource extended to those trains carrying the mail. Other trains, however, were largely defenseless. The IS was so effective that railcars carrying the mail were identified as such, and criminals would purposely avoid entering those cars of a train.

3 At the time, McLean headed a trucking company that bore his name but that was eventually taken over by CSX only to later be spun off. Most recently it was acquired by A. P. Moller Maersk (aka Maersk Lines).

4 Most railroads traditionally used 450 miles as the breakeven point for intermodal moves to become economically feasible. Recent initiatives have reduced this number, however.

Bibliography

Campbell, B. (2001). Marines as mail guards a story of the roaring 20's. *Military Trader*. Retrieved April 20, 2010 from www.therailroadpolice.com/

Crime Control Act of 1990, Section 1704. Title 49 Code of Federal Regulations. Railroad police authority.

Donzalski, D. (2009). Authority of the railroad police. *Police Chief*, 77:3.

Elliot, H., and Flake, B. (2010). An unlikely training partner: CSX Railroad. *Tactical Response*.

Gips, M. (2005). Teamsters bash rail security. *Security Management*.

Modafferi, P., and Bouche, K. (2005). Intelligence sharing: Efforts to develop fusion center intelligence standards. *Police Chief*, 72:2 (February 2005).

Piatt, B. (2007). *The improving American's Security Act*. http://dpc.senate.gov/dpcdocpr.cfm?doc_name=1b-110-1-34. Accessed February 2, 2009.

Saunders, R. (1978). *Railroad mergers and the coming of Conrail*. Westport, CT: Greenwood Press.

U.S. Government Accountability Office. (2009). *Federal rail security: Actions have been taken to enhance security, but the federal strategy can be strengthened and security efforts better monitored*. GAO-09–243.

5 Securing rail freight operations I

Securing the logistics chain (including intermodal)

During the Fall 2012 semester, an MBA course assigned the article "America's Most Logistics Friendly Cities"[1] and asked students to post comments on an online forum. Among other points, the article highlighted the importance of rail freight service, but several students posted comments including "Why are we bothering to discuss such an outdated and antiquated mode of transportation," and "Nobody seriously ships anything by rail anymore." This was surprising because all of the students had been born after the railroads were deregulated in the 1980s and, once again, railroad stocks were rated as growth stocks. Since economic deregulation, specifically through the enactment of the Staggers Act in 1980, freight railroads have been demonstrating an impressive and steady comeback. At first the railroads began by cutting back on unneeded infrastructure by single-tracking underutilized lines that previously were double tracked (and on some roads double-tracking formerly triple-tracked mainlines), and by selling off unprofitable branches, often to shortline and regional operators that were not burdened by high labor rates and extensive overhead costs. Illustrative of this comeback is the story of the Butler, NJ, resident who in 1982 opined that the railroad (in this case the New York, Susquehanna, and Western, or NYS&W) would never again be running freight trains through this suburban to rural community in the northwest part of the state. Two years later, that same resident was complaining about the traffic backups at grade crossings while NYS&W trains over a mile long traversed the newly renovated line hauling Sea-Land containers from Port Elizabeth, NJ, to Buffalo, NY. However, to gain an understanding of how to go about securing a freight railroad requires an understanding of the theory of supply chain.

The supply chain (or, as this chapter makes reference, the logistics chain) is the series of linkages of activities between buyers and sellers. In the late 1990s, the Supply Chain Council developed the Supply Chain Operations Reference (SCOR) Model that depicts the activities with each firm sequentially performing *source*, *make*, and *deliver* activities controlled by an overarching *plan* function. Each is a link, but the *source* activity connects the firm to its suppliers: the *deliver* activity, with its customers (Supply Chain Council). Moreover, every organization[2] is engaged in some form of *make* activity whether that is fundamentally changing the form of a product through manufacturing or assembly, or configuring it differently for potential use by a downstream customer – as when a retailer arranges product displays. Heading upstream, the supply chain therefore links a firm with its immediate suppliers such as the body panel stampers, tire producers, and interior components furnished to an automobile assembler. Those first-tier suppliers each have their suppliers, also thought to be second tier in terms of the automobile firm. The links continue backward up the chain until reaching such point that a basic material is either harvested from the land or extracted from the earth.[3] *Delivery* functions cease when products are consumed and finally returned to the earth even if that requires that they pass through multiple recyclings. The railroads can generally be thought to move basic materials, often in bulk, known as *carload business*, whereas finished products may more often be moved intermodally or by truck.

Figure 5.1 superimposes basic freight car types on a part of the SCOR Model. Whether it is raw material originating at a supply source, finished products en route to customers, or intermediate components moving between manufacturing plants, transportation is often described as "that necessary glue that holds the supply chain together."[4]

The SCOR Model does offer another dimension by specifically identifying the three basic flows: physical, information, and financial, which are shown in Figure 5.2. In the context of protecting railroad infrastructure, the focus is most often on the physical flow, because that is what is usually thought to be interrupted whenever there are attempts to destroy infrastructure. In some cases, those attempts are not just aimed at the immediate destruction of terminals, rights-of-way, or cargo but anticipate longer-term adverse effects on economic activity. Hence, it is the physical flow as represented by the ability of the railroad network to move materials and products from points of supply to points of demand. This means that it is the physical flow that enables the financial flows otherwise representing overall national economic activity. Therefore, it will be within the context of these three flows illustrated by the SCOR Model that will become the framework for considering protecting infrastructure for freight operations.[5]

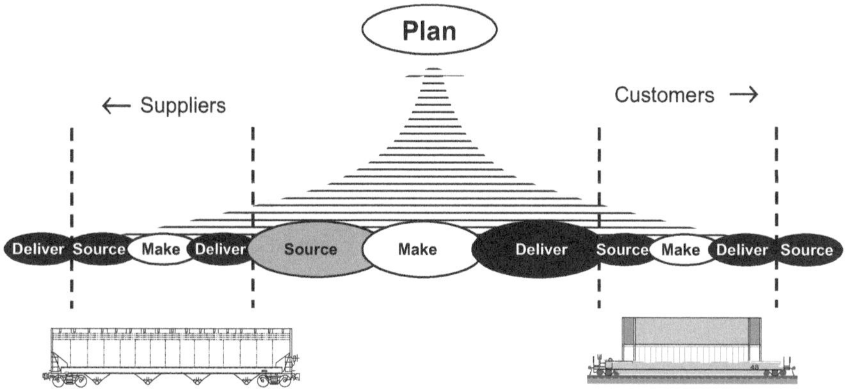

Figure 5.1 SCOR Model applied to railroad freight markets

Source: adapted from Supply Chain Council

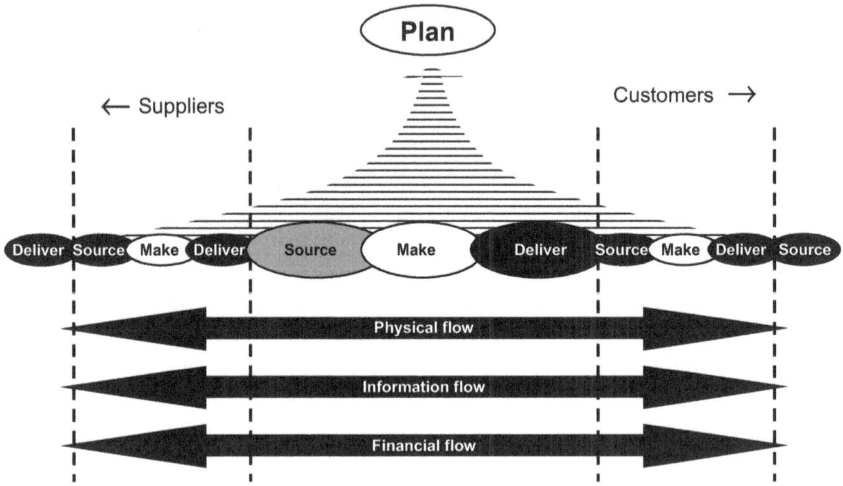

Figure 5.2 SCOR Model with flows illustrated

Source: © APICS 2015. APICS, CCOR, CPIM, CSCP, DCOR, SCOR, and SCORmark are all registered trademarks of APICS. All rights reserved.

Scenarios for freight movement

To the layperson, there are two just types of trains: freight and passenger. But freight movement needs to be further examined to understand the diversity of rail freight. Parsing freight movements further, there are several distinct categories that need to be taken into consideration:

- Unit trains of bulk commodities (i.e. coal and grain)
- Intermodal trains
- General merchandise trains
- Some combination of these (line haul)
- Switching and terminal activities

Each of these needs to be juxtaposed with four other criteria in order that the respective risks may be identified. These include:

- Origin, inclusive of specific shipper
- Destination, inclusive of specific consignee
- Routing
- Contents.[6]

Each of these criteria is discussed with particular consideration to the threats represented to the infrastructure, operations, and adjacent populations.

Unit trains

Perhaps the simplest to understand and to assess is the unit train that originates with a single shipper, in route to a single consignee, often in equipment owned or operated by the shipper, the consignee, or the railroad, and hauling a single commodity. A unit train is meant, by definition, to move intact with no intermediate switching or classifying of cars in order to send them in myriad directions. The most typical cargoes are coal, metallic ores, crude oil, grain, and construction materials such as gravel and sand; but, in wartime, military equipment may also fit this definition (Cavinato 1989). Increasingly there are intermodal unit trains moving between specific nodes. Some are dedicated to individual shippers, such as UPS.

Unit trains of a single commodity spend little time idle in yards and, with the exception of hazmat trains (perhaps crude oil and ethanol being key examples), generally make poor prospects as terrorist weapons. Another key factor is that their chain of custody is usually quite short. However, because a unit train may consist of 100 railcars or more, derailing one could disrupt a key rail line, but there would be no widespread destruction or significant loss of human life.[7] Consequently, with minimal hazards, the planned routings would focus on network capacity and the option of tying up a right-of-way with slower generally low-value freight. All of that said, during the period of time that we have been assembling this book, we have seen emerge yet a different twist on the use of unit trains: moving large volumes of crude oil from oil fields to refineries. See the sidebar on the movement of Bakken shale oil for a more detailed understanding of the dynamics of this growing business as well as some of the risks already seen and a discussion of the regulations that are emerging, in both the U.S. and Canada.

Intermodal[8]

Some intermodal trains are purchased in their entirety by trucking companies such as UPS, J.B. Hunt, or Schneider National; ocean container lines the likes of NYK or Maersk;[9] or intermodal service retailers, the major ones being Hub Group, Pacer Stacktrain, and Alliance Shippers. For the railroads this means that all of the cars share a common origin and destination pair – namely loading and unloading at intermodal terminals or directly in port areas. A 100-car intermodal train may have several hundred containers, often representing a large number of shippers and consignees as well as a large variety of contents. Most often those contents are generally considered to be of high value, containing consumer electronics and designer garments among others, and thus often making containers good targets for theft especially in rail yards, terminals, and wherever trains may sit idle or operate at low track speed. Some containers or truck bodies will be from less-than-truckload service providers, meaning that a single unit may carry a wide range of items for a variety of shipper-consignee permutations. The main issue of concern is the long chains of custody, but perhaps more important are the echelons of required documentation as well as the remoteness of the shipper from the railroad as the actual provider of carriage.

For illustrative purposes, consider the case of four different intermodal moves: (1) an ocean container of home electronics originating in China destined to the importer's distribution center in Chicago; (2) an ocean container of textile garments (t-shirts, athletic wear, and sweaters) loaded by a consolidator in Indonesia and shipped to a deconsolidation point that is located near Chicago; (3) a trailer of retail shipments loaded in New Hampshire and bound for individual consumers on the U.S. West Coast; and (4) a trailer of chemicals in drums that will move intact from Delaware to Oklahoma. Each of these is very different, but to the railroads they look very much the same in that they occupy the position of a loaded trailer or container on an intermodal flatcar.

Our first container was loaded at the manufacturer's plant somewhere inland in China, transported by highway to the Port of Shanghai, a port where it would be pre-screened by CBP before ever being loaded aboard a U.S.-bound vessel, where it remained for five days before being loaded onboard a Maersk ship destined to the Port of Los Angeles–Long Beach. Upon arrival, the container was offloaded from the ship, cleared U.S. Customs without even being opened, was then taken to the rail side of the container yard, and loaded aboard a double-stack train destined for Chicago. A couple of days later, the train arrives at a Chicago terminal where the container is taken from the train, mounted on a container chassis, and trucked to the consignee's distribution center. The total move entailed a shipper, two truckers, two ports, a ship line, a railroad, and a consignee. The container was closed and sealed by the shipper and was only opened by the ultimate consignee in the U.S.

The container of garments from Indonesia has a much more complex scenario. A myriad of small manufacturers each used a different small independent trucker to ship several cartons of product to an ocean freight consolidator[10] located at the small port serving that relatively remote area of this island nation. The consolidator identified those cartons that had to be loaded together, booked the shipment with the ocean carrier, stuffed the container, and transported it to dockside where it sat for three days awaiting the arrival of its assigned vessel. Thereafter it was loaded and taken to Jakarta, but only after making several intermediate port calls was it finally unloaded and again stored in a container yard. Eventually, it was loaded aboard a Neptune Orient Line ship and transported to Houston, where it was unloaded from the ship and opened, with its contents removed for careful inspection and clearance by CBP, before being re-stowed and braced by a cooper, and trucked to an intermodal terminal where it was loaded aboard a train destined for Chicago. A day and a half later, the train arrived, the container was offloaded and trucked to the de-consolidator where the goods were removed, and individual cartons were transported by several different truckers to their various retailer consignees.

In this example, the shipment required three truckers, two shipping lines, three container terminals, three ports where the container idled (but several more where it passed through), and two consolidators/de-consolidators. Additionally, there was a cooper plus a customs house broker involved with intermediate handling as well as arrival paperwork.

In the case of the trailer of retail shipments, an online seller has been taking orders on a 24/7 basis. At 7 o'clock in the evening, the trailer containing those packages, after being sorted for their addressees in the San Francisco Bay area, is picked up by a UPS tractor and driven to an intermodal terminal outside of Boston, where it is checked in and placed in a parking spot next to a waiting CSX intermodal train. Within a couple of hours, the trailer is loaded onto a flatcar, and the train departs for the Midwest. Not quite two days later, it arrives at an interchange point where the entire train is taken over by the Union Pacific that will take it onward to Oakland. There, the trailer is removed from the railcar and trucked to a sorting facility where the packages are sorted for local delivery in and around the Bay Area. It was a simple and straightforward move using a single trucker that had purchased the entire use of the train. While two railroads were used for the linehaul, UPS remained in charge, and the entire point-to-point move was executed in a seamless fashion. There were few points where the trailer idled and was thereby required to wait for the next leg of its journey.

The final example is a trailer that got loaded with 80 fiber drums of a specialty polymer that originated in Delaware. The shipper loaded and sealed the domestic container, which was then taken to a Norfolk Southern intermodal ramp outside of Philadelphia, removed from its chassis wheels, and loaded aboard an intermodal railcar. Not unlike the trailer of retail packages, the freight train headed west. At Kansas City, the trailer was offloaded, placed on a chassis, and positioned in a numbered parking spot on the terminal grounds. Several hours later a Schneider National tractor connected to the trailer and took it to its final destination, a specialty plastic molder. Since all goods were meant for a single customer and loaded by a single shipper, the trailer was under the control of either a single railroad or a single trucking company, which meant that the chain of custody was a simple one. Because the trucking companies and major container lines – again identifying UPS, but also J. B. Hunt, Schneider National, Maersk, and NYK Lines as perhaps some of the largest – purchase the use of entire trains and demand a particular level of service, the railroads assign a high priority to these, a noteworthy circumstance since it means that these trains seldom are idled waiting for higher-priority trains to pass.

The intermodal network has been the fastest growing segment of the railroad industry ever since it was deregulated with the passage of the Staggers Act in 1980. Table 5.1 lists the relative volume of intermodal freight that was moved in 2011. Intermodal technology has been a development that has helped with security by removing the exposure of break-bulk goods while en route and, hence, their vulnerability to theft, but it has increased security issues with respect to international goods because of the uncertainty of custody for those containers outside of U.S. soil. Since the events of 9/11, there have been several noteworthy initiatives to counteract these risks. These are discussed in detail later in this chapter.

Table 5.1 Significant U.S. intermodal routes (millions of annual net tons), 2011

Tonnage	*Between*	*And*	*Key intermediate points*[a]	*Railroad*
>20	Chicago	Los Angeles	Kansas City, Albuquerque	BNSF
10 to 20	Albuquerque	Amarillo, TX		BNSF
	Chicago	Salt Lake City	Omaha	UP
	Chicago	Cleveland		Multiple
	Dallas	Los Angeles	El Paso, Phoenix	Multiple
4 to 10	Albuquerque	El Paso		BNSF
	Amarillo, TX	Dallas		BNSF
	Amarillo, TX	Memphis, TN		Multiple
	Chicago	Seattle	Minneapolis	BNSF
	Cleveland[b]	New York City	Albany, Syracuse, Buffalo	CSX
	Cleveland[b]	New York City	Harrisburg, Pittsburgh	NS
	Dallas	Houston		Multiple
	Dallas	St. Louis, MO		Multiple
	Jacksonville, FL	Orlando, FL		CSX
	Los Angeles	Salt Lake City		UP
	Oakland, CA	Salt Lake City		UP
	Portland, OR	Salt Lake City		UP
	Portland, OR	Seattle		BNSF

Data sources: Federal Railroad Administration, *BNSF System Map 2014*

Notes

a Intermediate points may also represent significant originating and/or terminating TOFC and/or COFC tonnage.

b There are two routes between these origin and destination pairs.

General merchandise

General merchandise rail freight transportation has been around for as long as the railroads have been hauling freight. Also known as carload service, it describes the situation when trains will transport a mixed assortment of freight cars carrying a wide range of goods. Much of the freight hauled can be categorized as bulk volume, dimensional cargo,[11] or specific classifications of finished products, such as automobiles and construction equipment. Table 5.2 shows the typical loadings across the most common freight car types.

Merchandise freight trains will carry some combination of these cars – some randomly distributed throughout the length of the train, others in blocks when several are destined for the same consignee or freight yard, and yet others, such as those carrying hazmat, which are isolated from the crew or cars carrying reactive materials. A general characteristic of merchandise trains is that they move between major rail yards and individual cars may be resorted into different trains as they are routed from origin to destination. This sorting, known as classification, is often a time-consuming activity, meaning that a car may sit idle for a period of time before resuming its journey.

In the past 40 to 50 years, the railroads have become solely a carload business as they have shed the less-than-carload (LCL) services, most notably the disappearance of the Railway Express Agency,[12] which was the principal consolidator of LCL at the time. After its disbandment by the railroads, the LTL trucking

Table 5.2 Principal types of railroad freight cars and their loadings

Freight car type	*Subtype*	*Use or lading*
Tank cars	Non-pressurized: petroleum	Crude and refined petroleum products
	Non-pressurized: other	Bulk liquids: chemicals, clay slurries, agricultural products including corn syrup and edible oils
	Pressurized	Compressed gases, ammonia, chlorine, propane, liquefied petroleum gas
Boxcars	General purpose	
Covered hopper cars	Agriculture	Grains
	Chemicals	Plastic resins, urea, potash, soda ash, caustic soda
	Minerals	Salt, cement, glassmaking sand
Auto racks	Components	Automotive components for assembly
	Finished vehicles	Automobiles, trucks, farm equipment, construction equipment
Hopper cars (open)	Coal	Coal and steelmaking coke
	Ore	Primarily iron ore
	Other	Building materials, wood chips
Flatcars	General purpose	Coiled steel, other
	Specialized	Industrial equipment, military vehicles and weapon systems

Source: the authors

companies took over these services and subsequently began using TOFC services of the railroads between long-distance points.

Combination freight trains (line haul)

A combination freight train means that a train contains some combination of the previously described types, most often intermodal mixed with a variety of general merchandise cars. Such trains will typically require some classification activities and will be subject to many of the same security issues as described above for intermodal trains.

Freight trains (switching, but otherwise known as "locals")

Railroads require a labor-intensive activity at both the route origins and the destinations to either pick up loaded cars from shippers or deliver them to consignees. Accomplished by shorter trains moving at slower speeds, these are the trains that require frequent stops to pick up and drop off cars. Such trains are usually short distance and increasingly operated by shortline or regional railroads.

Vulnerability

Railroads face a variety of risks, and such risks can make railroad infrastructure as well as moving trains vulnerable to terrorist threats. The infrastructure is inherently vulnerable by the environment in which railroads operate: geography ranging from heavily populated urban locales to desolate remote deserts, mountainous terrain with a plethora of bridges and tunnels to elevated operations above city streets, all punctuated with those key nodes where freight is gathered or distributed.

Protecting the railroad infrastructure that carries freight trains means protecting it per se from those seeking to damage or destroy it, but recognizes that such may be inflicted by actions intended to damage trains. Said another way, destruction of a train can serve as the means to destroy infrastructure, which serves to potentially inflict physical harm on the population. Second, destruction of the infrastructure has the direct impact of disrupting commerce by breaking links in myriad supply chains – most freight trains carry a range of goods from raw materials to finished products heading to markets. In its own unique way, each represents a link between a buyer and a seller, or in SCOR terminology between a *deliver* function and a *source* function.

Geographic dispersion

It is instructive to consider that industry clusters, or those greater vicinities where companies from a specific industry tend to locate, have a distinct impact on railroad operations, specifically with regard to the commodities carried. Considering origins and destinations for major rail-borne commodities, there are several notable patterns that correlate well to specific freight car types and freight movements – some notable examples could include the CSX orange juice train, solid trains of crude oil, ethanol, mineral acid, and even grain.

Tank cars – non-hazardous cargo

A sizeable number of tank cars are employed in the movement of non-hazardous materials, most notably agricultural items such as vegetable oils and sweeteners and industrial slurries, the two major ones being a range of clays used in papermaking and titanium dioxide, which is a white pigment used in a variety of applications including but not limited to paint, plastics, and consumer products such as toothpaste. Most such cargoes are often transported in privately owned or operated tank cars by such firms as J.M. Huber, DuPont, ADM, Cargill, and A.E. Staley. A major destination for the clay slurries is the paper industry, which has numerous plants in New England, while consignees of the other materials are reasonably dispersed across the U.S. Other such clusters of consignees could include plastic molding resin to the greater Toledo area to satisfy the many molders of automotive components, or animal feed to central Pennsylvania for livestock production.

Finally, there is the matter of intermodal containers – the discussion can be bifurcated into trucking and ocean containers. There is some overlap in this discussion because the topic became blurred with the advent of the domestic containers during the 1990s. The rule of thumb for revenue-adequate intermodal moves is about 470 miles, recognizing that the costliest part of the service occurs with the loading and unloading of the trailers or containers from the railcars. While domestic intermodal moves continue on the rise, these have become commonplace with volumes clustered near population centers. In this section *international containers* are the topic of interest because of the volumes that are concentrated at the load center seaports, which can be segmented into Northeast, Southeast, Gulf Coast, and West Coast. Note, too, that there are inland ports located some distance from the obvious seaport, but where arrivals and departures of containers are coordinated with vessel sailings. Immediately coming to mind are Front Royal, VA, and Charlotte, NC, located on Norfolk Southern and CSX, respectively. As shown in Table 5.3, the largest ports are those where service historically began and developed. Several smaller ports are served by feeder vessels originating at the load centers, but as global container services grew in volume these increasingly have been served with direct calls by the major container lines.

Note that, in the above table, New York / New Jersey, Norfolk, Charleston, Savannah, Miami, Houston, Los Angeles / Long Beach, Oakland, and Seattle/ Tacoma are those with the longest legacies for international containerized traffic. Many of the connecting rail routes are still land bridge, meaning coast-to-coast,

Table 5.3 Major CONUS containerport traffic 2014

Rank	*Port*	*TEU's 2014*[13]	*TEU's 2015*
1	Los Angeles	8.34	8.16
2	Long Beach	6.82	7.19
3	New York–New Jersey	5.77	6.37
4	Savannah	3.35	3.74
5	Seattle-Tacoma	3.39	3.53
6	Hampton Roads	2.39	2.55
7	Oakland	2.39	2.28
8	Houston	1.95	2.13
9	Charleston	1.79	1.97
10	Port Everglades	1.01	1.06
11	Miami	0.88	1.01
12	Jacksonville	0.94	0.92
13	Baltimore	0.77	0.84
14	New Orleans	0.49	0.52
15	Philadelphia	0.45	0.43
16	Wilmington (DE)	0.33	0.34
17	Wilmington (NC)	0.28	0.29
18	Palm Beach	0.26	0.27
19	Boston	0.21	0.24
20	Mobile	0.24	0.23

Data source: American Association of Port Authorities

but there remains substantial traffic of a mini-land bridge nature, such as Los Angeles / Long Beach[14] or New York / New Jersey to Chicago.

Protecting actual physical flows

Given that most train accidents incur little significant direct damage[15] to infrastructure, consideration needs to be given to the externalities including the availability, expertise and cost of first responders, specifically fire, rescue, and medical services provided, the cost of rerouting trains perhaps on other railroads around the out-of-service affected area, and finally, the cost of disruption to customer supply chains.

Following the terrorist attacks of 2001, the AAR responded with the formation of an industry task force charged with developing a security plan and procedures[16] that built a database of all U.S. critical railroad assets, assessed the relative vulnerabilities of those assets to the continuity of railroad operations, and assessed the terrorism threat. The second phase of the AAR action was to assess the nature and severity of risk related to each of the vulnerabilities, determine what actions could be taken to thwart those risks, develop precise alert levels,[17] and define the actions that would need to be taken for each. Perhaps the crowning achievement, however, was the creation of an AAR incident operations center and its complementary alert system to which all major North American railroads were parties.

The public sector responded with the National Infrastructure Protection Program (NIPP), which in its 2013 revision articulated the objective of "managing the risks from significant threat and hazards to physical and cyber critical infrastructure through an integrated approach to identify, deter, detect, disrupt and prepare for threats; reduce vulnerabilities of critical assets, systems and networks; and mitigate the potential consequences to critical infrastructure of incidents or adverse effects."[18] While much of the focus of this book has been on terrorist threats, the NIPP strives to also encompass cyber threats, extreme weather, and accidents or technical failures (NIPP 2016).

Given the wide range of possible threats, the NIPP therefore was designed to embrace the greatest number of stakeholders, including industry, plus federal, state, local, and tribal governments. Industry, representing the private sector, however, is not a single entity. Considering the topic of rail freight would by definition need to also include other participants in the supply chain besides the railroads, such as shippers, consignees, and other modes of transportation, especially in light of intermodalism, to include the motor carriers, drayage providers, and marine container lines.

Bringing critical infrastructure protection closer to the potential sources of the problem requires the active participation of subnational jurisdictions as well as federal agencies. Many have established statewide critical infrastructure protection plans. In the case of Pennsylvania, this involves collaboration between the Pennsylvania Emergency Management Agency, the State Police, and the Office of Homeland Security. California, in contrast, has taken a more aggressive stance in having its Public Utility Commission personnel gain access to and be able to comment on the security plans of the Class I railroads (note that there are only two: Union Pacific and Burlington Northern Santa Fe) operating in the state.

Each Class I railroad as well as many of the regional and shortline railroads[19] have developed their own tailored security plans that are based on the AAR model and that are reviewed with TSA.[20] That said, however, RAND conducted research on railroad vulnerability and concluded:

> We conclude that much of the infrastructure is accessible to would-be attackers and is vulnerable. However, the consequences due to attacks are relatively minor. Safety incidents in rail freight generally do not result in casualties and do not incur significant direct damages to track or equipment. Incidents involving the release of hazardous materials, however, can have far more severe consequences. Were such an incident to be perpetrated with the intent of maximizing casualties, the results would be catastrophic.
>
> (Ortiz et al. 2008, p. xxi)

Given that railroads are the preferred transportation mode for hazardous materials (hazmat) and that the implications for not only the railroad but the public at large are so great, the topic of protecting freight railroads moving such materials has been given its own separate chapter. The movement of oil by rail is also significant enough – and highly charged politically given recent incidents – to warrant separate treatment in a following chapter.

Protecting information flows

Where this book may take its most significant departure from much current thinking is in its approach to information flows. We began the chapter with a discussion of the SCOR Model and how its three flows of physical, information, and financials were intertwined, but perhaps it is most important to consider that it is the information flow that is the enabler of the other two. The physical movement of goods has historically been dependent upon information in the form of purchase orders, sales orders, bills of lading, packing slips, waybills, customs declarations, and invoices to provide the instructions for what the economists would label *gaining the utilities of time, place, quantity, customer, and price*. In the age of information technology, many of the formerly paper documents are now available electronically. Initially, such information was shared internally on legacy systems, but in the past 15 years legacy systems, clearly for transaction-related data, have given way to enterprise resource planning (ERP) systems, the best known probably being SAP.[21] For the past 30+ years, firms have sought to increase connectivity, especially between buyers and sellers, and have begun applying electronic data interchange (EDI) systems.

While the first commercial applications of information technology was the automation of commercial systems, even in the 1960s the railroads began to apply potential solutions to their operational issues. Railroads, not unlike many of the parallels found in commercial aviation, rely on communications and signaling systems to control train movements, but they also have adopted control systems for aligning track, detecting train defects such as hotboxes and dragging equipment, and providing the location of locomotives and freight cars on their networks using radio frequency identification systems (RFID).[22] CTC systems allow railroads to control train speed and direction remotely but also enable them to operate more tonnage over fewer track miles by more efficiently and systematically managing train movements. They have become as a result a major bonus to productivity. Moreover, CTC has allowed the railroads to shrink their infrastructure and staffing requirements so as to accommodate increases in volume without needing to make the capital investment in more track. More recently, the FRA has mandated that PTC systems be installed on all road locomotives[23] to reduce the potential for error that could result in derailments and potential loss of life should two trains occupy the same track. Originally installed in 2000 on locomotives operating on Northeast Corridor where Amtrak passenger trains, Norfolk Southern and CSX freight trains, and an assortment of commuter agencies operate daily, PTC has enjoyed an excellent safety record. PTC, however, is a concept rather than a single system, given that there are several providers offering different levels of sophistication, but nevertheless all are compatible and offer the same general benefits.

Cyber attacks regularly make the news most often involving the banking system, whether it is compromising credit card numbers from major retailers or shutting down the transaction systems of a major bank. Most firms are aware of the vulnerability of their accounting systems, and, if they aren't, soon will be. Railroads certainly also have transaction systems that log sales, track railcars, and bill shippers, but their vulnerability is primarily with those systems controlling train movement, whether these are the signaling systems or those that align switches within their networks.

Admittedly, information technology has developed into a double-edged sword. With all of the improvements to productivity and safety comes the security issue of, what are the implications if these systems are hacked? Would the lowest level of attack be a hacking into PTC to permit two trains to occupy the same track at the same time, thereby causing a collision resulting in a derailment? Or does the opportunity exist for a hacker to deactivate all of the signaling aspects so that they show as "clear" or "green" and cause derailments that would tie up an entire railroad with major interruptions to the U.S. economy? Much like the topic of hazmat, a separate chapter is provided later in this book to address the issue of systems as it relates to infrastructure utilized by railroads.

Protection

It has been established that the freight railroads are vulnerable. This is not due to a single factor: there is a wide range of potential threats, some of which are more ominous than the others. In discussing threats from terrorism to rail freight, it is important to keep in mind the usual goals of terrorists, to disrupt normal life, gain publicity for catastrophic events, cause significant loss of life or physical damage, and generally disrupt the economy. The SCOR Model is, perhaps, a useful template for looking at threats and vulnerabilities. The traditional threats would all be focused on the physical flows because they are the most visible to both the potential perpetrator as well as the public at large. An attack against the infrastructure has its publicity value that can best be described as shock and awe. However, derailments of freight trains that do not involve hazmat[24] result in relatively little damage and are seldom newsworthy. A previous report from RAND stated that attacks on the rail system are most likely with the use of small hand-carried devices mainly because the logistics of being able to place larger ones becomes exponentially more complex to execute (Wilson et al. 2007). Moreover, the disruption may be an inconvenience, but in most cases freight trains can be rerouted either over the target railroad's lines or those of others. Small-device attacks could have greater impact should they be carried out in a coordinated fashion that would target multiple critical points on the network simultaneously. RAND further concluded that access to rail lines is a critical feature, meaning that subterranean lines are less vulnerable while elevated ones are more so. For the freight railroads, idling engines and trains as well as yards and loading/unloading infrastructure pose the greatest vulnerability while track, bridges, and tunnels pose less (Ortiz et al. 2008). We show threat and vulnerability as juxtaposed in thc matrix in Figure 5.3, but make note that the concept of the coordinated multiple attacks has not been taken into consideration.

For physical protection, guarding against trespassing would appear to be a key solution. Increased use of surveillance equipment is a frequently employed tactic, which can be achieved through increased vigilance as well as technology means. In Chapter 3, we made reference to the use of railroad police, including the intelligence efforts stood up by some of the more progressive units. While rail security still requires some physical presence by officers, the use of remote monitoring has been on the increase, and as this is written there has been discussion about the potential use of drone aircraft to monitor rail lines and yards.

	Vulnerabilities			
Threats	Idling locomotives and trains	Yard & loading/ unloading equipment	Bridges and tunnels	Track and right-of-way
Small explosives	***Higher***			
Large explosives & incendiaries				
Unconventional weapons				***Lower***

Figure 5.3 Vulnerabilities and threats

Source: Adapted from RAND 2008

Two other initiatives that are notable include the previously mentioned Citizens for Rail Security Program instituted by the Burlington Northern Santa Fe that to our thinking establishes a railroad equivalent of the Neighborhood Watch, but has now been adopted by others, most notably Amtrak and Norfolk Southern. Clearly, the advantages realized by BNSF have been noticed by others in the industry. Of equal note is the program to certify contractors that may be working on railroad property; whether these are maintaining systems or rights-of-way, such certified workers also become additional knowledgeable observers.

Although somewhat minimized by earlier assessments of threat and vulnerabilities to rights-of-way, the focus needs to shift to the threats and vulnerabilities to the information flows employed by the freight railroads. Recent attempts by hackers, some believed to be state-sponsored, into myriad systems utilized by military and commercial organizations, suggests that there may be vulnerability to those train control systems employed by the railroads. Earlier research suggested that communications and signaling systems could be breached with the result being significant disruption, not of just a single train or a single line, but of the entire railroad network (Plant and Young 2007). With the U.S, being principally served by seven Class I railroads, each employing robust information systems, there may well be significant vulnerability. While requiring much more technical expertise than that required for physical attacks, the potential for detection, especially when compared to that required for multiple coordinated physical attacks, is extremely low. Moreover, the magnitude of the potential damage, especially to the economic activity of the nation, is quite real.

The specific approaches to protecting information technology are a topic beyond the scope or objectives of this book. There is a growing volume of literature that addresses how information systems can be protected from attacks, whether these stem from internal or external sources. The rail industry needs to stay abreast of such developments and work closely with government agencies concerned with information security and protection.

Summary

Freight railroads carry significant quantities of bulk raw materials, especially those used in the energy and agricultural sectors. They move finished goods loaded into containers and trailers as intermodal freight and automobiles in specially designed auto carriers. Disrupting these flows can be achieved through a direct attack on the infrastructure enabling the physical flows, or through an indirect attack on the information technology that enables those physical flows. To this end, the SCOR Model is a useful template or lens for considering the interaction of these flows.

At the time of this writing, no significant incidents have been reported involving the interruption of freight movement by terrorists, or major threats to indicate this form of transport is a likely target. However, the threats from terrorism are constantly changing, and today's sense of security may easily be broken by a single incident. Most chilling to contemplate is the use of freight rail as a means to transport weapons of mass destruction, perhaps in an intermodal container that has not been adequately checked at some point in the supply chain, or in a grain car that has been contaminated to spread disease or infect livestock. Or, in a different scenario, an invasion of a railroad's signal system could cause a massive collision between a 10,000-ton freight train and a passenger train sharing trackage. The range of risks and vulnerabilities are vast and without empirical evidence hard to assess risk.

Protecting freight infrastructure relies upon established proven human assets as well as qualified interested parties whether they are railroad contractors or rail fans. Technology has its role in protecting the physical flows, but vigilance, including intelligence gathering, will prove useful for protecting the information flows that are the key enablers of the physical flows.

The nature of private railcar fleets in North America

A little-known fact is that the majority of railcars in North America are owned and/or operated by firms that are NOT in the railroad or even in the transportation business. In one case, TTX (also known as Trailer Train Corporation), such a firm is a joint venture between several railroads and owns the majority of the intermodal equipment, specifically the double-stack units, spine cars, and flatcars used for transporting trailers and/or containers. The other railcar providers are typically employing covered hopper cars, tank cars, and assorted open hopper cars for the movement of such materials as grain, fertilizer, coal and coke, ore, aggregates, petroleum, plastic pellets, and chemicals.

Owning versus operating

Non-railroad organizations owning rail cars are divided into leasing companies and firms that produce, sell, or use bulk commodities. Among the leasing companies, the major names include GATX, GE Leasing, CIT Leasing, and several of the major banks. The commodities firms may elect to lease rail cars or own them outright. Of the operators, one will find most of the major chemical companies such as Shell, Huntsman, BASF, Dow, DuPont, and Exxon. Agricultural products companies, also major fleet operators, include Archer-Daniels-Midland (ADM), A. E. Staley, Cargill, Continental Grain, and the Canadian Wheat Board. Finally, those firms using open hopper cars would include Pennsylvania Power and Light, Detroit Edison, Peabody Coal, Joseph, and Brandenburg.

How this came about

Two major factors came together to promote the use of private rail cars: (1) the chemical producers began using cars as *warehouses on wheels* where they would store the finished product awaiting customer orders or as a means of maintaining consistent plant throughput by evening out the ups and downs of the demand cycles, and (2) the need for cars with increasingly specialized characteristics such as lining systems, unique alloys for tank vessels, and valve and piping arrangements. Chemical producers do not like the extra expense of cleaning cars between shipments, nor do they want to deal with the troublesome issue of residual quantities remaining in cars. As a consequence, tank cars were moving loaded in only one direction and returning empty. Thus, they were seeing a significant reduction in loaded miles traveled, which was to curtail the annual revenue that each could generate.

All were factors that led to more expensive cars with lower utilization factors thus driving down the returns on investment, which was a critical factor for the railroads as many were barely profitable and by the 1950s some were already in bankruptcy.

Rise of the leasing companies

Manufacturing companies sought to avoid investment in railcars, given that car ownership, and with it fleet management and ongoing car maintenance, was not one of their core competencies, and they began leasing cars from the car builders. With time the finance companies also entered the market. Today some of the leading leasing firms include Trinity Industries, ACF, and Union Tank Car among the builders, while General Electric Finance holds the distinction of being the world's single largest railcar owner, a position that becomes even more impressive considering that they are also a major lessor of a wide range of transportation equipment including locomotives, ocean containers, highway trailers, and aircraft (The Official Railway Equipment Register).

During the late 1980s came the advent of the third-party logistics provider (3PL) as an outgrowth of firms in the trucking, warehousing, leasing, and information services industries. Their aim was to bundle a range of services tailored for a specific firm, thereby leveraging logistics-related core competencies that few manufacturing firms could claim similar credit for. Several firms leasing private railcars bundled car inspection and repairs, transloading, and bulk storage.

Today, the railroads own virtually no tank cars or covered hopper cars for hauling crude oil, chemicals, or plastic resin.[25] Where one does find such equipment, it is typically for hauling non-hazardous commodities, most often for agriculture.

Customs-Trade Partnership Against Terrorism and the rail link in the chain of custody

Not long after the attacks of 9/11, U.S. Customs (later Customs and Border Protection, or CBP) rolled out an initiative, the *Customs-Trade Partnership Against Terrorism*, better known as *C-TPAT* for short.[26] The underlying premise was that importers needed to be able to verify the efficacy of their global inbound supply chains by obtaining and maintaining first-hand knowledge of the security for those goods sourced from foreign suppliers to the delivery at destination. Many major firms were early to sign onto the initiative because the quid pro quo was the more expeditious movement of arriving goods through the Customs clearance process – primarily because the goods destined to these importers were deemed to be of relatively lower risk. Goods of importers that were not C-TPAT participants would therefore take longer for clearance, perhaps several days longer. A companion program, the Container Security Initiative (CSI), involved stationing CBP agents at major foreign ports of loading to vet U.S.-bound containers before they were ever loaded aboard a ship heading to the U.S. In both cases, CBP sought to significantly reduce the possibility that threatening materials could ever reach U.S. soil. Note that goods can also arrive in the U.S. by rail, either as stuffed into intermodal containers or loaded aboard conventional railcars. Today, C-TPAT has over 10,000 member firms including importers, carriers of all modes, and customs house brokers (www.cbp.gov).

C-TPAT was a game changer for many importers because it meant that they had to have knowledge of their supply chains that in the case of goods by sea could potentially include myriad participants. Moreover, the respective INCOTERMS[27] that identified which party was paying for what services and where title passed had little impact because if a firm was the importer of record, it was the responsible party. Consider the following example:

A U.S. chemical manufacturer in Missouri purchases a specific compound[28] from a supplier near Frankfurt, Germany. Being only 12 drums, the purchase quantity is less than a full container; hence, the supplier engages a consolidator[29] to pick up the drums and load them into the next available container heading to the U.S. The full container is trucked to an inland port on the Rhine River, where it is loaded aboard a barge bound for Rotterdam, where it will be offloaded, trucked to the container terminal, and loaded aboard a Maersk ship destined to the U.S. Upon arrival in Baltimore, the container is removed from the ship and taken to a customs inspection area where its contents are checked. The container is then restowed[30] and a new seal installed. It is subsequently placed on an intermodal railcar that is part of a train heading for St. Louis. At St. Louis the container is taken to a container station where it is stripped of its contents; our 12 drums of chemicals clear customs and are given to a trucker for the final leg of the journey.

In this example, there are 14 participants that either have custody of the shipment or, as in the case of the customs broker, have knowledge of it, as is shown in the following list in Table 5.4:In this example, there is only one commodity purchased from a single supplier at a single location and ultimately delivered to one destination, but in reality firms that engage in significant importing may have dozens of U.S. locations requiring dozens of suppliers providing hundreds or even thousands of products. The key point is that each may represent a separate supply chain that needs to be documented and verified for the purposes of C-TPAT.

In retrospect, the process for vetting supply chains has meant a radical departure for global supply chain management at firms, including many that are among the largest and best-known names. These needed their procurement organizations to actually visit suppliers and accumulate and document substantial information about ownership, management, security measures, and their own respective supply chains. In many instances, these visits were the first time that many suppliers were ever seen, but it was only one step in the verification process that would ultimately lead to C-TPAT acceptance – the culminating step would be a CBP review of the importers' processes, followed by on-site visitation to the suppliers.

Table 5.4 Participating entities in an international intermodal move

European side	*United States side*
Supplier	Baltimore container terminal
Consolidator	CSX Railroad
German trucker	St. Louis trucker
River container terminal	Deconsolidator
German barge line	Customs broker
Dutch container terminal	A second St. Louis trucker
Maersk Line	Importer

For the railroads,[31] C-TPAT membership had a twofold effect. First, it offered added value to customers by being able to verify the chain of custody for imported goods. In the case of containers, these are inspected and/or cleared at the seaports of arrival; in the case of conventional railcars inspections occur at the rail yards in Mexico before departing for the U.S.; for cars arriving from Canada, at the first rail yard in the U.S. Secondly, it gives the railroads a better understanding of what they are transporting and for whom – not total confidence, but a much improved picture when compared to pre-9/11 operations.

Source: https://www.cbp.gov/border-security/ports-entry/cargo-security/c-tpat-customs-trade-partnership-against-terrorism. Accessed October 12, 2014.

Notes

1 Bill King, *Modern Materials Handling*, November 19, 2007.
2 To emphasize this point, the author once engaged a USMC officer and described how a maintenance company did produce something when it turned a vehicle in need of repair into a functioning one.
3 Consider iron ore extracted from an open pit mine and corn harvested. However, there is one possible exception here: that being the liquefaction process of separating oxygen, nitrogen, hydrogen, and miscellaneous rarer gases from air.
4 Dr. Richard Young uses this description in his graduate transportation courses.
5 Note, too, that protecting freight rail infrastructure will often mean protection of the passenger services that may be tenants of the freight railroads and utilize the same tracks and other infrastructure.
6 Hazardous materials require a detailed discussion that is covered in a separate chapter.
7 Transport of crude oil, such as in the case of the Lac-Mégantic incident, would be an exception.
8 Also known as TOFC (for trailer on flat car) and COFC (for container on flat car), even if contemporary railroad equipment does not meet the conventional description of a flatcar.
9 The concept of the ocean container dates to the late 1950s when a trucker, Malcolm McLean, who headed a firm bearing his name, saw an opportunity for providing faster loading and unloading of ships by keeping goods intact within a trailer body. That first service was instituted with the name Sea-Land Service and became the standard by which all containerized ocean freight remains transported today.
10 In the maritime trades, these parties are known as non-vessel operating common carriers, or NVOCCs for short.
11 Dimensional cargo is also known as "heavy, high, and wide", which well defines its characteristics. Large machinery and structural steel are good examples.
12 Also known at REA, the Railway Express Agency was not a railroad but rather owned by the railroads on a joint venture basis. It was disbanded in 1967.
13 Twenty Foot Equivalent Unit
14 The best example may be the Alameda Corridor, which is used for staging export containers as well as moving import containers 20 miles away from the Port of Los Angeles–Long Beach.
15 Note that some accidents are catastrophic, such as the crude oil carrying unit train at Lac-Mégantic, Quebec in 2013.

16 Note that the details of this security plan is not available to the public because it is security sensitive information (SSI).
17 The AAR's four levels of alert are level 1: normal day-to-day operations; level 2: heightened awareness; level 3: credible threat of attack on the railroad industry; and level 4: confirmed attack against the railroad industry.
18 See also the sidebar for a more detailed discussion of the NCIPP.
19 Shortline and regional railroads have continued to consolidate with the result being firms holding portfolios of these smaller railroads. Notable examples are Genesee & Wyoming Corporation; WATCO; Pinsley Railroads; Pioneer Railcorp; and the Livonia, Avon, and Lakeville.
20 Note that the members of the ASLRRA also adhere to a similar protocol.
21 SAP is the name of the enterprise resource planning system best known by its purveyor named SAP. SAP systems seek to integrate manufacturing, sales, human resources, finance, and supply chain management into a common system.
22 The earliest approach to car identification employed a clerk standing trackside writing down the reporting marks on the cars in a train as it passed by. Later, barcodes were applied to freight cars and read by electronic scanners but proved unreliable when the coded placards became covered with road grime. By comparison, RFID provides automatic equipment identification (AEI) in all weather, day or night, and at any speed all with high levels of reliability.
23 The FRA deadline for installing these systems was December 31, 2015. Switching locomotives limited to yard service are excluded under the most recent federal rules.
24 The topic of hazardous materials transportation is covered in depth in the next chapter.
25 The soon-to-be exception will be when BNSF takes delivery of 5,000 new DOT 117 tank cars for transporting Bakken crude oil. The reader is directed to Chapter 6 for a more detailed discussion on this topic.
26 CBP knew that the largest importers, representing a significant portion of imported goods, had many of the processes in place or, if not, the resources to provide them. Moreover, CBP was aware of the value of product velocity and its financial impact on inventories. C-TPAT would hence allow CBP to direct its limited resources to areas where import processes are not so well developed.
27 INCOTERMS or *IN*ternational *CO*mmercial *TERMS* are the standards established by the International Chamber of Commerce.
28 For simplicity this example will have chemicals that are non-hazardous.
29 For ocean cargo, this party is known as a non-vessel operating common carrier, or NVOCC.
30 In this example the container will not clear customs in Baltimore but travel in bond to St. Louis.
31 This is the case for all of the U.S., Canadian, and Mexican Class I railroads but includes many of the regional lines as well.

Bibliography

Cavinato, J. (1989). *Transportation-logistics dictionary* (3rd ed.). Washington, DC: International Thomson Transport Press.

National Infrastructure Protection Program. (2016). *U.S. department of homeland security*. Retrieved January 5, 2017 from www.dhs.gov/national-infrastructure-protection-plan

Ortiz, D., Weatherford, B., Greenburg, M., and Ecola, L. (2008). *Improving safety and security of freight and passenger rail in Pennsylvania*. Santa Monica, CA: RAND Corporation.

Plant, J., and Young, R. (2007). *Assessment of terrorist threats to the U.S. railway Network*. Boston: Citizens for Rail Safety.

Wilson, J., Jackson, B., Eisman, M., Steinberg, P., and Riley, K. J. (2007). *Securing America's passenger-rail systems*. Santa Monica, CA: RAND Corporation.

6 Securing rail freight operations II
Hazardous materials

On a cold December morning, a group of businesspeople await the arrival of their 7:12 a.m. commuter train that will take them to yet another day at work. As they clutch their steaming cups of coffee or perhaps text colleagues from their smartphones, a local freight train rumbles past on the far track much like it has on so many other mornings – this scenario is quite routine. On this particular morning the freight train is hauling a string of generally nondescript black tank cars with yellow numbers and lettering. None of the commuters pay any attention to the fact that some of them bear red diamond placards lettered "flammable" while others carried black-and-white diamond placards lettered "corrosive." The fact that each tank car bears four letter reporting marks[1] ending in "X" was not only of no concern to the bystanders on the station platform, but not a one knows what that means.

To the group of commuters, trains such as this one were commonplace. Moreover, if any had taken the time to read some of the other markings on these freight cars, they could have likely elicited other information, most of which would probably be meaningless as casual passersby would have had little inkling of what *vinyl acetate monomer* or *sodium hydroxide* happened to be, what they might be used for, or that they might pose a serious health hazard if the substance happened to be released. The fact that they gave this train, in general, and the freight cars, in particular, no unusual notice is a tribute to the success of the railroads in moving large quantities of hazardous materials with very few incidents.

Clearly, the movement of hazardous materials has been commonplace on North American railroads almost since the time that they first began carrying any quantities of freight. This chapter discusses some of the history of why this occurred, the nature of today's transport of hazardous materials, and some of the regulations governing such movement, and it explores some of the issues of when and why this topic needs to be interwoven with that of safety, terrorism, and the protection of infrastructure.

The potential

Perhaps the blasé attitude towards tank cars of hazardous materials rolling nearby is the result of the industry's impressive safety record, but the potential for disaster is nevertheless great. Flammable liquids, especially when present in volume, can wreak widespread devastation, even to the point of burning down entire towns.[2]

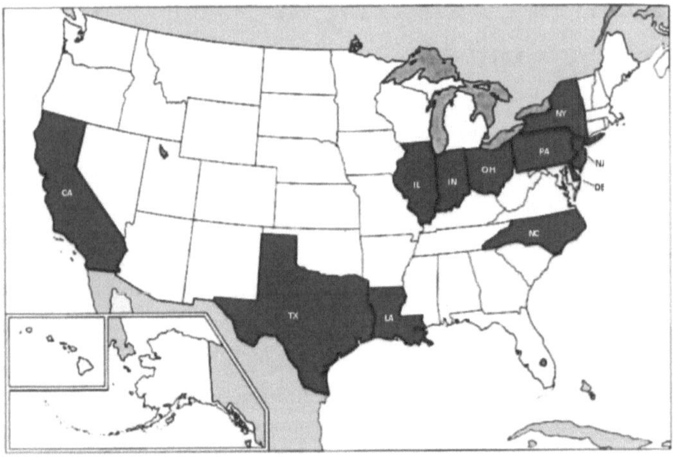

Figure 6.1 Ten major chemical-producing U.S. states

Data source: Association of American Railroads

The ignited contents of a single ruptured tank car can spread to adjacent ones. Poisonous fumes can kill residents as well as inflict harm to flora and fauna – ammonia and chlorine being the most prevalent such toxic materials, moving in prodigious volumes by rail. With the combinations of chemicals carried by a single train also possible, a multi-car derailment may have the potential of combining materials to yield a range of deadly agents. Figure 6.1 shows that ten U.S. states account for approximately two-thirds of the U.S. chemical production. While these represent the origins of chemical shipments, their use is widely dispersed and could include papermaking in Maine, synthetic fiber production in North Carolina and South Carolina, plastics manufacture in Ohio and Michigan, and agricultural chemicals formulated for use in the Midwest.

Hazmat and railroads: growing up together

The first industrial revolution in the early nineteenth century moved manufacturing from a highly fragmented cottage industry to one where factories not only employed large groups of workers, but one that could achieve significant economies of scale. What some have described as the second industrial revolution, or those 45 years between 1870 and 1914, was a period of invention that saw the advent of the petroleum, steelmaking, agricultural, and chemical industries (Mokyr 1998). It is not a coincidence that the railroads, first established in the U.S. in Baltimore in 1830, were key enablers of the growth of other industries.

The U.S. petroleum industry began in Titusville, PA, in 1859; steelmaking grew due to the invention and early introduction of the Bessemer Converter and the large-scale coke works that were built in and around Pittsburgh; and the chemical industry emerged from the ability to produce chlorine and caustic soda from common salt as was the case of Dow Chemical in Michigan. Other developments soon added even more materials needing to be moved in bulk. Agricultural productivity was boosted with the addition of chemical fertilizers, most notably nitrates, phosphates, and potassium. These industries, however, had lots of common ground with other manufactured commodities. For example, coke works also created volumes of coal tar as a byproduct that yielded asphalt and also benzene for chemical use. Petroleum cracking provided not only a range of fuels and lubricants but also the basic chemical building blocks of benzene, toluene, and xylene. Where the railroads fit into the equation was through their ability to move very large volumes of basic materials from the point of production to the point of consumption of these products.

In the late 1800s and early 1900s, given that roads (1) were unsuitable for long distance travel, (2) were incapable of handling heavy loads, and (3) had low travel speeds, there were only two viable modes of transportation: inland waterways and railroads. The former were severely constrained by weather conditions and access considerations, all issues that the railroads, as a mode of improved transportation,[3] easily could overcome. To do so, however, required the development of the tank car and covered hopper car to accommodate liquid bulk and dry bulk commodities, respectively. Largely evolved during the past 150 years, these two types of railcars today comprise a major portion of the North American railcar fleet.

As the fleet has technologically developed and grown, so has the variety of materials transported. For agriculture, fertilizers were augmented with crop protection materials, specifically in the form of herbicides, fungicides, and insecticides. The chemical industry continued to blossom to include such common and general-use plastics as polyvinyl chloride, polystyrene, polyethylene, polyester, and polypropylene, to name just a few. The railroads were increasingly called upon to move large quantities of raw materials in the form of basic chemical building blocks, intermediates specifically as the monomers[4] for producing polyvinyl acetate, polypropylene, styrene monomer, polyester, and nylon, as well as finished products.

As materials have become more complex, so have the number of hazardous materials needed to be transported. However, hazardous materials do not necessarily mean that the products manufactured from them are hazardous. For example, polymers needed for producing automotive components, packaging materials, and a vast array of consumer goods are non-hazardous despite their precursors variously having flammable, poisonous, corrosive, or toxic-by inhalation properties. In the twentieth century, chlorine became the ubiquitous disinfectant of municipal water supplies; caustic soda had among its major uses metal cleaning and industrial water softening. Large volumes of chemicals and fuels, all having one or more hazardous characteristics, were preferably transported by rail. Even with the development of the nation's highway system, rail transport continued to be seen as the safest mode for moving these items in volume.

In the twenty-first century, in response to political decisions against building more pipelines to connect newly discovered oilfield deposits with refining capacity, the producers of crude oil turned to the railroads as the next-best approach for moving the large volumes. It was not long before the oil companies also realized the inherent flexibility of the railroads, not limited to specific routing like pipelines, whereby supply could more easily meet changing patterns of demand. Long freight trains of crude oil have become commonplace in both the U.S. and Canada. This development has not been without incident and controversy. Several major derailments have caused explosions and fires, often with devastating impact. The worst occurred in the small town of Lac-Mégantic, Quebec, in 2013. An improperly braked train accidentally rolled into several parked tank cars of liquefied natural gas, setting off an explosion and igniting the crude oil with devastating results that included 47 deaths and extensive property damage. (See the following chapter for an extensive discussion on the transport of crude oil and ethanol.) Later that same year, there were other crude oil train accidents in Louisiana and West Virginia that made national headlines, but fortunately it resulted in no loss of life, even if it did cause major environmental contamination to nearby wetlands.

Current practices

The movement of hazmat by rail can generally be focused into just a few categories: tank cars, covered hopper cars, and drums or intermediate bulk containers (IBC)[5] in boxcars. Intermodally, hazmat is transported as liquid in ISO tanks[6] and as either liquids or solids in drums and IBCs loaded into domestic or ocean containers mounted on flatcars.

As capacities for chemical production have increased, innovation advanced in the industry with a rapid expansion in the number and types of materials produced. For the movement of large volumes of chemicals, even with the advent of the modern highway system, the railroads remain the preferred mode of transport. With the growth of the chemical industry came a demand for railcars having a diverse range of characteristics that included tank sizes, tank vessels of different metals, a variety of tank lining materials, and different piping and valves for the control of loading and unloading. Chemical producers began to insist on cars being assigned to dedicated service in order to avoid expensive cleaning that also resulted in the collection and disposition of any residues in an environmentally responsible manner. Cars in dedicated service also precluded the potential for cross-contamination of product, but their use also meant that the railroads would need to make major investments in freight cars that would have lower utilization factors, hence lower returns on invested capital. The unanticipated consequence was that the increasingly cash-strapped railroads passed the responsibility and costs back to the shippers and, in some situations, to the customers for chemicals. In other words, the railroads will gladly transport tank cars and covered hoppers of chemicals and plastics loaded into those cars owned by or leased by industry.

Today, nearly half of the North American freight car fleet is privately owned, much of it by the chemical and related industries in fleets of tank cars and covered hoppers.[7] Table 6.1 shows a sampling of some of the larger operators of private rail car fleets among key firms in the chemical industry.[8] Rail car design is governed by standards established by the AAR in conjunction with the Federal

Railroad Administration. In the case of those transporting hazmat, there are specific requirements for non-overriding couplers,[9] non-vulnerably located piping appurtenances,[10] reinforced bulkheads,[11] and double-walled vessels.

U.S. railroads, as shown in Table 6.2, transported more than 30 million carloads of freight in 2011. Of this total, slightly more than 6 percent are of hazardous chemicals, including those classified as toxic by inhalation (TIH), which approximate 0.3 percent of total carloads. The related safety record shows 99.997 percent of these reached their destination without a release caused by a rail-related accident. Most shipments of chemical products by rail do not represent hazardous

Table 6.1 Major private railcar fleets in North America[a]

Company	*Tank car fleet*	*Covered hopper fleet*	*Products transported*
BP Chemicals	3,100	4,950	Polymers, monomers, BTX[b]
BASF	573	1,124	Polymers, caustic soda, chlorine
Celanese	1,815	813	Polymers, acid, methanol
Conoco Philips (chemicals)	1,178	7,172	Polymers, monomers, BTX
Dow	5,359	4,706	Polymers, glycols, solvents, other
DuPont	2,584	1,549	Polymers, acids, other
ExxonMobil (chemicals)	5,820	10,031	Polymers, monomers, BTX
Procor	13,863	4,097	Gases, other
TOTALS	33,520	34445	

Data source: Official Railway Equipment Register

Notes

a Fleets registered to chemical producers only. Note that actual fleets may be larger due to the number of additional units leased from such firms as First Union, GATX and General Electric Leasing.

b BTX, or benzene, toluene, and xylene as byproducts of oil refining and basic chemical building blocks.

Table 6.2 Class I chemical carloadings and tonnage originated

Year	*Originated carloads (mio)*	*Originated tonnage (mio)*
2003	1.90	161
2004	1.99	168
2005	1.94	165
2006	1.95	167
2007	2.05	176
2008	2.04	175
2009	1.90	162
2010	2.21	186
2011	2.27	192
2012	2.05	171

Data source: Association of American Railroads

materials. For the railroads, chemicals, both hazardous and non-hazardous, were the second-most prevalent lading by tonnage behind coal, as shown in Table 6.3.

For intermodal equipment, the vast majority is owned by TrailerTrain Corporation,[12] a joint venture between the Class I railroads in North America. Initially, intermodal cars were specially modified flatcars designed for carrying highway trailers, but the technology eventually evolved with lightweight and efficient articulated spine cars for moving highway trailers as well as containers mounted on chassis, and articulated double-stack well car sets for international and domestic containers. Efficiency derived from the absence of the container chassis[13] that would otherwise add weight, but also saved on height thereby enabling the stacking of two containers. Due to clearance restrictions, not all rail lines can accommodate double stacks, but railroads and on occasion jurisdictions have invested in infrastructure improvements to allow their movement on much of the nation's rail network.

Finally, most of the railcars used for the transport of chemicals are not just vehicles for the movement of material but also an important source of storage capacity. Chemical processing plants generally do not have a large range of operating throughputs but rather are run at or near capacity or are otherwise idled. Producers, therefore, use rail cars as portable storage units. In recessionary times, what would ordinarily be vacant spurs in and around major chemical manufacturing areas, with Houston being a notable example, are instead occupied with loaded tank and covered hopper cars. The key point is that these cars are stored in a vast range of isolated remote areas, often without the slightest oversight or physical protection.

Railcars containing hazmat are just not confined to tracks adjacent to their respective producing plants, but they can be found in equally isolated and remote points near customers' plants, on spurs awaiting switching into customers' plants, in rail yards, both large and small, and in intermodal transfer facilities. The latter are terminals where loaded rail cars are consigned waiting to have their contents

Table 6.3 Class I commodities by tonnage and revenue

Commodity	*By tonnage*	*By revenue*
Coal	43%	22%
Chemicals	10%	13%
Non-food agricultural	8%	9%
Non-metallic minerals	7%	4%
Food	6%	8%
Metallic ores	4%	1%
Primary metal products	3%	4%
Misc. mixed shipments*	6%	13%
Forest products	3%	5%
Motor vehicles	1%	7%
Other	9%	16%

Data source: Association of American Railroads

* *Includes intermodal*

transferred to trucks for transport to the final customer. Historically, rail cars in such locations were largely ignored, but recent years have been subject to vandalism and theft. Vandals at the least leave graffiti, but with increasing frequency they are releasing car brakes and/or opening discharge connections. It is a great leap of faith to consider such cars as unlikely terrorist targets, a topic that will be discussed later in this chapter.

Containers and truck bodies being transported intermodally are typically not used as storage facilities and, when mounted on flatcars or double-stack cars, are not idled outside of plant perimeters. Where they do tend to be stationary is in container yards, waiting to be loaded or having just been unloaded from rail cars.

Mindful that substantial liability is attached to the transport of hazmat, the railroads do place limitations on what may be moved and how. While hazmat represents only 5.2 percent of rail traffic, TIH is a mere 0.25 percent. The railroads would be happy to decline such business, but under their common carrier status are not permitted to do so. In recent years, nearly 20 percent of the chemicals in North America are moved by rail, with quantities doubling since 1980, a pivotal year for the rail industry when it was economically deregulated.[14] How the industry got to this point needs some explanation (AAR 2011).

Some regulatory history

The regulations regarding the movement of hazardous materials by rail have grown, as the public and their representatives in Congress grew increasingly aware of the possible dangers involved in freight rail transport. Regulations have been developed for both the type of equipment utilized and the practices involved in moving various cargoes.

While there were rail car standards during the first half of the twentieth century, they were largely promulgated by the railroad industry, specifically the AAR. Moreover, during that time, most of the technical expertise resided with the engineering departments of the major railroads themselves, but increasingly they were influenced by their industry counterparts, namely the major chemical producers themselves.

The 1970s saw a near perfect storm emerge with respect to moving hazmat. First, the railroads were increasingly going bankrupt, merging with other railroads, or both. Increased availability of reliable truck competition – thanks to the nearly completed Eisenhower Interstate Highway System and an Interstate Commerce Commission that was intent at keeping the railroads tightly controlled – strained many railroad finances, but perhaps none more than those in the Northeast, where most, if not all, were in bankruptcy or close to it. A shortage of cash was also manifested in deferred maintenance practices that often resulted in derailments with hazmat cars not being exempted. The railroad industry's claim that it was the safest mode of moving hazardous materials was in jeopardy of becoming a thing of the past.

Congress, intent on reigning in the potential for disaster, passed the Hazardous Materials Transportation Act of 1975, which addressed the practices of all of the modes, including the railroads. As codified in 49 CFR 173, the responsibility for safe transportation is a joint effort that includes the shippers, or as stated in the regulations, "those offering goods for commerce." In doing so, shippers were required do the following after determining that a particular material met the definition of a hazmat:

(1) Ensure that the proper shipping name is used. Some chemicals can be designated by several names, hence using a single common one assures that first responders, in the event of an incident, correctly understand that with which they are dealing and know how to both protect themselves as well as correctly engage in cleanup activities.
(2) Make certain that the hazard warning label is correct. Materials need to be properly labeled, which means not only having the name of the material correct, but its hazardous properties as well.
(3) Ensure that the correct packaging is used. Packaging, per the regulations, means not only small packages, but larger units such as drums and bags. 49 CFR 173.24 stipulates specific types of packages and limits them by mode and quantity.
(4) Provide timely training for those employees who not only directly handle specific hazmat, but those with responsibility for documentation and for arranging transportation.
(5) Provide correct shipping documentation for the hazardous shipment.
(6) Furnish correct information for emergency response. There is actually a twofold element to this that includes the responsibility of the shipper itself, but also contact information for Chemtrec©,[15] the hazardous material information center operated by the chemical industry group, the American Chemical Council (ACC), formerly known as the Chemical Manufacturers Association, where expertise is available to carriers and first responders for a range of hazardous materials events.
(7) Make certain that the vehicle carrying the hazmat, whether a truck, barge, or railcar is displaying the correct external placards, as stipulated by law, to alert first responders.[16]
(8) Have a safety plan in place for handling mishaps. Note that in some cases, the shippers have their own hazmat response teams or may have a contractor fulfilling this role. The major firms also engage in mutual aid agreements to provide better geographic coverage as well as additional resources in the event of a mishap of major proportions.
(9) Provide timely incident reports. The Hazardous Materials Transportation Act requires that shippers compile and regularly file with the U.S. Department of Transportation documentation that discloses any accidents including location, transport mode, name of shipper, name of carrier, and other particulars.

Generally, the regulations place the same responsibilities on both shipper and carrier. The carrier needs to verify that the actions of the shipper are consistent with the regulations, but also have its own security plan in place.

Although the Hazardous Materials Transportation Control Act of 1970 (49 USC §§ 5101–5127) was the first piece of legislation, and for many constitutes the cornerstone regulation, it was considerably strengthened by the Hazardous

Materials Transportation Act of 1975 that contains many of the responsibilities placed on shippers and carriers. The act has been reauthorized in the Hazardous Materials Transportation Uniform Safety Act of 1990 and the Hazardous Materials Transportation Authorization Act of 1994.

Other regulations have been promulgated that address the transportation of hazardous waste, which often has seen the railroads as the preferred mode of transportation. These include the Resource Conservation and Recovery Act (known as RCRA, 90 Stat 2795), the Comprehensive Environmental Response, Compensation, and Liability Act (also known as CERCLA, 85 Stat 431–461), the Superfund Amendments and Reauthorization Act (SARA), and its various amendments and reauthorizations.

With over 2,700 substances of regulatory interest, it is the transportation of explosives, gases, flammable liquids and solids, oxidizers, poisons, radioactive materials, corrosives, and hazardous waste that makes for such an onerous task for shippers and railroads to comply. Nevertheless, the safety record of the railroads in this regard is enviable, with 99.997 percent of all carloads reaching their destinations without incident (AAR 2011). The railroads have achieved an impressive 88 percent reduction in releases between 1980 and 2007.[17] Admittedly, the bulk of this improvement was achieved during those first ten years after economic deregulation with the passage of the Staggers Rail Act in 1980, during which time the railroads were able to make major investments in improved track conditions, new technology for communications and signaling, and reliable second generation locomotive power. This was also a period of time when weaker (and usually less well-maintained) railroads could not compete and were abandoned or spun off lines to shortline and regional roads. The application of technology solutions for improved safety has also included investments in dragging equipment and hot journal[18] detectors[19]; digital track geometry measurement equipment; periodic rail grinding, tie replacement, and ballast cleaning; and PTC, although the latter has not been fully implemented as of mid-2017.

Safety and security

There are two forms of regulation that share elements in common but that also may result in unintended confusion and conflict. Railcar design standards meant to reduce the potential for unintended release in the event of a derailment serve both safety and security concerns. However, the placarding of railcars for the purpose of alerting first responders to hazardous properties also serves to notify terrorists and others with ill intent of the whereabouts of such hazardous materials.

While a major aspect of regulating safety addresses "protecting property and people from the hazardous material," the point of regulating security is, perhaps, twofold: (1) protecting the integrity of the hazardous material from people vis-à-vis contamination or potential theft and (2) protecting the hazardous material from being illicitly used by those intent on doing harm to either property or others. Hence, safety and security can be deemed two sides of the same coin. This point is raised because much of the means of achieving safety has a direct application to the ability to provide security.

Moving hazardous materials safely and efficiently has been the focus during much of the history of the railroads. During two world wars, protection of the entire range of railroad infrastructure was a challenge that was met by the joint efforts of the war department and the railroads themselves. Hazmat were given little thought, as it was believed the threats to the rail network lay in its ability to move materiel in great volume from the inland factories, where it was produced to the ports of embarkation where it was subsequently loaded on ships bound for the theaters of war. The same logic applied to the movement of troops to theaters of war.

Fast forward to the new millennium, when commercial aircraft were turned into weapons to destroy buildings and kill civilians. Not only were railroads seen as transporters of goods to consumers, but they were seen as the means of conveying harmful materials that could maim and kill civilians. The transportation picture changed overnight as a newly established Department of Homeland Security attempted to find ways to protect infrastructure and the citizenry, and the TSA was formed with the charter to protect all modes of transportation. TSA, however, only concerns itself with the secure custody of railcars when stationary and for the tracking of such cars while in transit, thus leaving the actual planning of risk-based routing decisions with other agencies as the FRA and the Pipeline and Hazardous Materials Safety Administration (PHMSA) within the U.S. Department of Transportation.

With much of TSA's emphasis and resources devoted to securing air transportation, the clear potential for rail safety and security issues has led many locales to pass their own legislation seeking to regulate the movement of hazmat by rail. Many believe that the impetus started during that summer just prior to 9/11, when a CSX freight train derailed and caught fire in a tunnel in Baltimore. Intense heat from the fire and the derailment of 60 cars made containment efforts difficult. Cities across the country sought to take individual action to prevent similar catastrophes. In 2005, another catalytic event was the derailment of a Norfolk Southern Railway freight train in South Carolina that led to the release of toxic chlorine fumes. Within a matter of months, Baltimore, Boston, Chicago, Cleveland, and the District of Columbia all had passed legislation banning hazardous materials shipments through their city limits when destined for other locales. While the courts have struck down all of these local laws, acknowledging that only the federal government can pass legislation affecting interstate commerce, more importantly their efforts to ban hazardous cargoes on some routes, if enacted, would have meant the shutdown of some municipal water systems due to a lack of the necessary chlorine for their treatment plants. The most significant development to date has been new standards for chlorine tank cars that are more crashworthy and less likely to release their contents in the event of a derailment or other incident.

Tension between safety and security

All vehicles – whether domestic or ocean containers, trucks, or railcars carrying hazardous materials – are required by law to display placards that disclose the nature of the cargo being transported. Within 49 CFR, regulations also require that such placards bear the United Nation (UN) four-digit number that describes the specific product.[20] Moreover, when the cargo is known to be a *toxic inhalation hazard* (TIH), that specific wording must also appear in large letters on the side of the vehicle. In the case of rail cars when in dedicated service, the specific name of the product will often be shown as well.

State and local emergency management agencies want to know what is being transported through their jurisdictions, by what route, and in what quantity. At stake is the need for first responders to know what they potentially need to deal with in the event of an accident and how to proceed. There are other perspectives on this issue, however, not the least of these being the sheer amount of data that might need to be compiled for some jurisdictions, such as Houston, where there are thousands of rail cars moving with myriad substances. Second, the question is raised, are the emergency responders capable of managing the quantity of data that would likely be made available in real time? Finally, there is the security issue of making publicly available the location and car identification for materials that could potentially be of terrorist interest.

Although previously posited that safety and security share many of the same concerns and interests, the identification of vehicles and their contents does pose some risk. There are some in industry who are concerned that even the placards may be disclosing too much information given that anyone can readily access the UN numbers and be able to readily target a specific rail car. Clearly, each represents a weapon of sorts that needs only to be breached to cause destruction of property, loss of life, and harm to economic activity. This problem is more compelling considering the joint use of rights-of-way such as that portrayed in the opening paragraphs of this chapter. With terrorist groups finding it easy to share information on such technical issues as bomb-making, use of 3-D copiers to make weapons, and other topics of potential harm, it is not fanciful to think that they would share information on placarding of rail cars. Perhaps they already have.

Addressing hazmat security

The chemical industry, railroads, and government have done much to advance hazmat transportation safety. Safety and security share some of the same characteristics but also have some marked differences including some that put them diametrically opposed. Hazmat shipments have been made substantially safer despite some inherent obstacles, but many of the security issues remain or have grown in importance in recent years. The chemical industry has done an admirable job by establishing Chemtrec under the auspices of the American Chemical Council, and the larger chemical producers have stood up their own emergency response teams that often have such a high level of expertise that they regularly train public sector first responders.

The major security risk is the difficulty in protecting the expanse of track structure, stations, and terminals through which hazardous materials move or are stored. The major chemical producers have generally attended to the protection of their plants, with some perhaps drawing best practices from the steel companies, most of whom had increasingly sophisticated protection processes that stem from the labor unrest of the late 1800s and much of the 1900s, but also the wartime threats to a critical defense-related industry. The exposure comes from those railcars sitting idle in rail yards, awaiting transloading services, spotted on unprotected consignee spurs, stopped at interlockings en route, shunted to passing sidings, or stored loaded in any manner of places – but also on slow-moving freight trains. Providing sufficient fencing, lighting, police patrols, and even high-tech solutions would prove to be a daunting task given the highly dispersed nature of the numerous locations involved.

The theft of materials from rail cars is not a likely problem, but the release of their contents is, whether or not it is a violent explosion attracting major media coverage and immediate widespread destruction of property and personal injury or death, or a slower leaking of poisonous substances. Other measures to isolate hazardous materials from places where they might be most destructive include rerouting where feasible. For some years the railroads have sought to avoid moving trains containing hazardous substances near large public gatherings such as sporting events. These are a matter of timing as well as location.

Reducing the number of hazardous items or for that matter their quantities is not a realistic option given the demands of industry in modern society. Changing transportation modes from rail to truck is not a viable alternative either. While trucks typically are not idled for nearly as long as the average railcar, the safety considerations of highway transport pose significant – albeit different, and arguably greater – risks.

Improving information sharing and management is an option, whereby notification of hazmat shipments can be made available to regional fusion centers rather than the myriad individual county and municipal jurisdictions along any given rail route.

The net outcome is that there is considerable security exposure due to the inherent nature of the U.S. rail network. The policing activities of the railroads, as discussed in a previous chapter, does an excellent job for what it is; however, that effort is not sufficient for the scope of the problem. From a security perspective, government regulation by TSA is not that agency's primary focus, nor does that agency have the resources to address the volumes of rail freight carried daily by North American railroads. The next chapter deals with a more recent and related phenomenon, specifically the transportation of flammable materials in large volume. Beginning in the late 1990s, ethanol has been extensively transported by rail, and more recently, with the discovery of readily extractable crude oil in the Bakken Basin in North Dakota, so has a particularly volatile type of crude oil. As flammable materials, these perhaps pose a greater risk than chemicals, but they do so for clearly attributable reasons. The following chapter covers this subject in further depth because it is a specialized and timely issue. Even as the volume of traffic involved has declined, public interest in safety and environmental concerns due to well-publicized derailments has kept the issue of the safe and secure movement of oil by rail a "hot topic," to use a bad choice of words.

Notes

1 Reporting marks indicate the name of the owner of the freight car, and those ending in "X" mean that the car is owned by a firm that is not a railroad, in this case either a chemical producer or a leasing company.
2 This refers to the events at Lac-Mégantic, Quebec, which will be discussed at length in the following chapter.
3 *Improved transportation* is the term used by logisticians to denote movement by mechanically propelled vehicles rather than human or animal powered.
4 Monomers are the basic chemical precursor that with the presence of a catalyst form long chemical chains better known as polymers.
5 IBCs are typically forklift-sized units that depending on design can accommodate liquid or dry bulk commodities. IBCs, constructed variously of metal or plastic, are returnable units.
6 ISO tanks are vessels mounted in a steel cage and are equal in size to a 20-foot ocean container because they were originally designed to be carried onboard a containership. Note, too, that domestic ISO tanks have become commonplace.
7 Cars may be owned by manufacturing companies, but they are also extensively leased from such firms as GATX, North American Car, Pullman National, and General Electric.
8 Railcars for petroleum products, specifically fuels and lubricants, have been excluded.
9 Coupler design that prevents the coupler of one car from riding up and puncturing the vessel of an adjacent car in the event of a derailment.
10 Valves and piping protected from damage in the event of a derailment.
11 Reinforced protection to minimize the potential for tank rupture.
12 It was founded in 1955, and over the years its ownership by the major railroads has shifted as the result of mergers, acquisitions, and bankruptcies.
13 A chassis is the trailer undercarriage that includes the wheel bogies, braking system, and landing gear necessary for highway operation.
14 The Staggers Act is the landmark legislation that unshackled the railroad industry and enabled it to reassume a growth trajectory that had been absent for decades.
15 Chemtrek© is staffed with qualified experts on a 24/7 basis and is funded by the member firms of the ACC. In the case of chlorine, the Chlorine Institute operates a parallel initiative, Chlorep©.
16 This issue also has the drawback of alerting a potential terrorist. This problem will be further addressed in more detail later in this chapter.
17 Note that most of this improvement came immediately following deregulation, and only a 39 percent improvement was achieved since 1990.
18 *Journals* is the term applied to wheel bearings; the term *hot journal* in the vernacular is *hot box* meaning an overheated bearing.
19 Automated equipment that can measure overheated journals to warn of possible wheel failure and resulting derailment.
20 A common product is UN 1203, denoting unleaded gasoline.

Bibliography

Abel, Z. T. (2011). Getting hazmat transportation back on track: The need for hazmat liability reform for rail carriers. *William and Mary Environmental Law and Policy Review*, 35(6), 973.

Association of American Railroads. (2011, March). *Hazmat transport by rail: An unfair liability*. Retrieved August 27, 2016 from www. thehill.com/sites/default/files/aar_ haz matbyrailseptember2009_0.pdf

Edwards, F. L., and Goodrich, D. C. (2013). *Introduction to transportation security*. New York: CRC Press.

Federal Register. (2008). Hazardous materials: enhancing rail transportation safety and security for hazardous materials shipments. *49 CFR 172, 174, and 209.*

Mokyr, J. (1998). *The second industrial revolution: 1870–1914*. Unpublished essay. Chicago, IL: Northwestern University. Accessed October 4, 2012.

Railroads and chemicals. (2013). *Association of American railroads*. Retrieved November 1, 2015 from www.aar.org/keyissues/Documents/Background-Papers/Railroads-and-Chemicals.pdf.

Railroads and ethanol. (2012). *Association of American railroads*. Retrieved November 1, 2015 from www.aar.org/keyissues/Documents/Background-Papers/Railroad-Ethanol.pdf.

Spraggins, H. B. (2010). The case for rail transportation of hazardous materials. *Journal of Management and Marketing Research*, 3(1), 1–8.

U.S. Office of Technology Assessment. (1986). *Transportation of hazardous materials*. NTIS: PB87–100319.

7 Securing rail freight operations III

Hazmat and the specific case of crude oil and ethanol

Most people are oblivious to freight trains that include tank cars loaded with various hazmat such as flammables, corrosives, and toxic inhalation hazards. While quite commonplace in those days prior to mid-2013, that apparent disinterest soon changed with the catastrophic events that occurred at a sleepy and otherwise unknown southern Quebec hamlet known as Lac-Mégantic. Soon the print media as well as television labeled trains hauling flammable materials, in particular those carrying crude oil from the Bakken region of Montana and North Dakota, as *bomb trains*.[1]

While the Lac-Mégantic disaster could have been avoided, one needs to consider the circumstances that contributed to its occurrence as well as understand the various actions that the industry[2] in concert with government is currently taking. Some of these have implications linked to security of the rail network, but at this juncture the premise must be made that the safety issue would be nowhere near as high a profile if the Lac-Mégantic incident had never happened.

Lac-Mégantic in perspective

In mid-July 2013 a train operated by the Montreal, Maine and Atlantic Railroad (MMA),[3] a company technically classified as a regional railroad, was en route on one of the last legs of its journey from New Town, North Dakota, to Saint John, New Brunswick, with a unit train of 72 tank cars carrying Bakken crude oil. The crew, consisting of only an engineer, stopped the train, set the brakes on only two of the cars and the five locomotives, and left the train. The plan was for a relief engineer to arrive and take over the train, taking it further eastward. Sometime later, in the absence of any MMA crew, the local fire department responded to a fire on the lead locomotive, extinguished the blaze, and shut down the unit's power, which also had the unintended consequence of shutting down the air compressor that kept the train's brakes applied. The train began to roll, slowly at first, but then picked up speed reaching what officials estimated to be in excess of 60 miles per hour. It eventually derailed, but did so where it punctured a car loaded with propane that was spotted on an adjacent track. The resulting explosion soon engulfed other propane cars, and the fire soon spread to the nearby tank cars carrying crude. Before the fire was fully extinguished, 47 people had died, scores of others were injured, and 2,000 persons had been evacuated. Moreover, half of the town was burned to the ground, but much of the remainder had been contaminated with crude oil and was required to be subsequently leveled[4] (Canadian Transport Safety Board 2014).

Underlying issues

As unfortunate as the incident at Lac-Mégantic was, it was, in fact, the coming together of a series of circumstances that if taken individually probably would not have produced a disaster of such magnitude. It was clearly the result of the converging issues of (1) the legal definition of crude oil, (2) the operating practices of the railroad, (3) the volume of material involved being significantly greater given that unit trains have a greater concentration of potentially problematic material focused in a limited space, and (4) the technical standards of the tank cars assigned to haul crude oil.

Under U.S. federal regulations,[5] crude oil is required to be carried in placarded vehicles displaying a red diamond containing the number "1267". However, that regulation treats all crude oil as if it was a single commodity, which if based on a commonly found variety known as West Texas Intermediate (WTI) has a Reid Vapor Pressure Index (RVP)[6] ranging from 5.0 to 7.0. Vapors, specifically as volatile gases, especially as butane, ethane, methane, and propane, are highly flammable and in sufficient quantity explosive. Bakken crude has a RVP of 12 to 14, which by comparison is just less than refined gasoline with an RVP of 15. The specifications of the DOT 111 standard would have been adequate if WTI was involved, but they were not sufficient for the Bakken loads (CARS 2016; Park 2014). Clearly, crude oil is not a homogeneous commodity, but, depending on its source, it has substantially different properties.

The train at Lac-Mégantic would likely have represented 15,000 gross tons, far too much weight for it to be held in place by just the brakes of two cars and the five locomotives. Prudent operating practices would have had the brakes on many more, but probably not all the cars also set.

Multiple tank cars of the same commodity, especially when compared to individual cars, pose a greater risk.[7] Such risk is multiplied by several-fold when an entire train consists of tank cars carrying a single commodity.[8] The potential for disaster could be expected to rise exponentially when one considers that a single tank car carries up to 30,000 gallons; a 20-car cut, approximately 600,000 gallons; and a 100-car unit train, 3 million gallons. Risk differs should 20 cars be connected as a cut as opposed to having those cars distributed at various points throughout the train. The key is the concentration of material within a given contiguous space or if such cars of perhaps different materials, but nevertheless of the same hazard type, are combined. An example of this would be such combinations as crude oil, liquefied petroleum gas, propane, alcohols, and various monomers used in the production of plastics (Plant and Young 2015).

Over the course of more than a century, the safety features of railroad tank cars have steadily improved. Those first wooden vats mounted on flatcars became steel vessels on flatcars, but then they evolved into the monocoque[9] design seen today. In the late nineteenth century, knuckle couplers became standard, and more recently the non-overriding double shelf coupler was developed to prevent couplers from piercing the vessels of adjacent tank cars in the event of derailment.[10] Those cars that derailed at Lac-Mégantic were built between 1980 and 2012 and complied with the FRA's DOT 111 standard,[11] which was likely adequate for equipment carrying crude oil if it had been WTI (Canadian Transport Safety Board 2014).

The point has been made that Lac-Mégantic was a terrible catastrophic event that should never have happened. It became that singular media event that raised the awareness of the public that hazmat were rolling through their communities. Moreover, it was occurring with a frequency where only few realized the potential implications – as illustrated by those commuters nonchalantly waiting for their morning train as a local freight rumbled past. While the railroads had traversed the landscape for nearly two centuries, the concern was now one of what they were carrying and what the associated risk was. During the past three years, other railroad accidents involving tank cars carrying crude oil have derailed, although none with the impact of Lac-Mégantic, and all have elicited extensive media coverage. A television station in Chicago ran a special dubbed "Bomb Trains", and activists have regularly protested such trains on the anniversary of Lac-Mégantic. In other instances, groups have protested such trains when it has become known that they pass nearby urban housing areas.

The topic of ethanol is included with the discussion of transporting crude oil for two primary reasons: (1) it is placarded as a flammable,[12] and (2) it is transported in multiple car units, either as cuts or as unit trains.

Public policy

The public policy response has been a multifaceted one and all with the realization that modern economies require the safe transportation of substances that present some elements of risk. The emergency response literature often speaks about the approaches of protection, dispersal, hardening, redundancy, and recovery, which is a useful manner in which to weigh the actions taken by both industry and government (Bush 2003).

Protecting: the *key train* concept

The FRA has identified trains in which it has a special interest because of the material(s) being carried, the quantities involved, and the tank cars being employed. Key trains are subject to lower speed limits, specifically 50 miles per hour, and will require route risk assessment, a topic that will be discussed later in this chapter. The attributes of a key train, as per FRA regulations (49 CFR 172) are:

- 20 or more cars of crude oil or ethanol, whether as a unit train, a single cut of cars, or an aggregate of 20 dispersed throughout the train, but also intermodal cars carrying T/PIH, flammable gases, or environmentally sensitive materials;
- five or more cars containing toxic/poisonous by inhalation substances, such as ammonia or chlorine;

- just one car of spent nuclear fuel; or
- any train of just one old specification DOT 111 tank car carrying a hazardous material.[13]

Hardening: providing a more resilient fleet of tank cars

The first response was hardening, or in the case of the railroad tank car, devising new specifications that prevented violation of the vessels and release of hazardous contents. The new cars, dubbed the "DOT 117 standard", were developed through the efforts of the AAR, PHMSA, and the FRA. The design includes several key elements:

- Tank vessels made from a thicker (0.5625 inches) steel
- Thermal jacks to better insulate contents from heat
- Protective full height head shields on each end (0.50 inches thick)
- Improved and protected bottom discharge valves
- Better protected top fittings to reduce damage in a rollover (Van Ness Feldman 2015)

While these improvements are valuable, clearly needed, and appropriate, the problem lies with the magnitude of the task of equipping a sufficient number of cars to handle the volume of freight needing to be moved. The North American fleet of tank cars is approximately 272,000 units. Many of these do not carry flammable hazardous materials and instead are relegated to transport everything from clay slurry to sweeteners such as corn syrup and molasses to nitric and sulfuric acids.[14] Of the total, 190,000 cars are intended to carry hazardous materials with 93,000 being for flammables. Of those, 40,000 cars are in dedicated ethanol service, which also represents a hazard being that it, too, is flammable. The remaining 53,000 cars are for all other flammable materials of which crude oil is only one commodity (AAR 2015). In short, there was a shortage of rail capacity, given the requirement for the transport of ethanol with its increased blending in gasoline, the advent of Bakken crude production, and the reticence of the federal government to permit the building of additional pipelines. Figure 7.1 shows the rapid increase in the movement of both ethanol and crude oil by rail and underscores the capacity problem vis-à-vis tank car availability.

With the pressure on the industry to begin using the newly specified cars, the order books of the car builders, such as Trinity Industries, GATX, and others, filled quickly. Moreover, those firms specializing in retrofitting existing cars saw their backlogs grow as well. By early 2015, the *Wall Street Journal* (WSJ) estimated that it would be well into the next decade before the nation's fleet was compliant with the DOT 117 standards.

(Figure 7.2 shows the growth and subsequent decline in new tank car orders.) At the same time, the *Wall Street Journal*, stating that the cost to retrofit was approximately the same cost for building a new car, estimated the total tab to exceed $6 billion in current dollars.

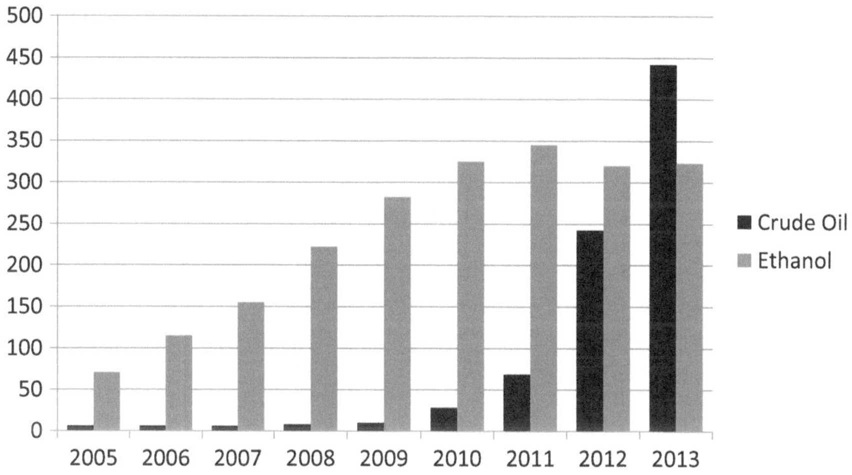

Figure 7.1 Annual rail carloads of crude and ethanol (thousands)

Data sources: Association of American Railroads and National Transportation Safety Board

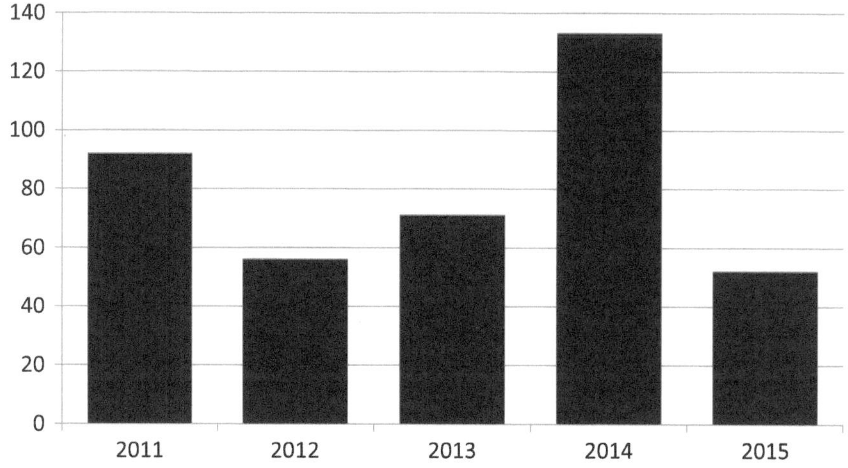

Figure 7.2 Orders for new tank cars (thousands)

Data source: Railway Supply Institute

In an earlier chapter, it was stated that the U.S. (and by inference much of the Canadian) fleet of freight cars is owned and/or operated by firms that are not railroads – in particular banks, leasing companies, and oil companies. The sole exception may wind up being BNSF, a unit of Warren Buffett's Berkshire Hathaway, which has just ordered 5,000 cars (Vantuono 2014). What makes this significant is that there is no single entity responsible for the end-to-end movement of a hazmat shipment. The implication here is not that the investment is not warranted, but rather the time required for implementation and the number of parties involved considerably raises the level of complexity.

Recovery: disclosing volumes and routes

In 2014 the Federal Railroad Administration issued an emergency order directing U.S. railroads to disclose the movement of significant crude oil and ethanol carrying trains to respective state agencies with the intent that (1) this information would be shared with the first responder agencies within those jurisdictions where trains would pass and (2) those agencies would be better able to provide effective emergency response should such need arise. Many such agencies proactively engaged the railroads and have established mutual drill opportunities for training personnel.

The FRA order also intended that such disclosures be sensitive security information (SSI). While many states honored this, several (including New York and Washington) took the position that such information was important to put into the hands of the public at large. While Washington reached agreement with BNSF, the railroad with the single most track within that state's borders, New York sought to post this information on a website. There are several situations where the railroads have refused to disclose train routings and negotiations have continued (Gordon and Young 2015).

Clearly, the railroads sought to comply with the emergency order, noting its purposeful intent, but they were equally concerned that any release of such information that details the volume, location, and timing of such trains (in light of the current security concerns) in the wrong hands could be used to plan terrorist activities with far-reaching consequences. Most states were able to reach a practical accord. This matter underscores how safety and security are topics that share many common interests but may also have reason to be at odds.

Dispersal: rerouting hazmat trains

Another approach with the potential to make a short-term impact is to minimize the potential effect of a derailment-caused material release. FRA mandates that railroads seek alternative routes for hazmat-bearing trains in order that major concentrations of population, major civic venues such as stadiums with NFL games, important historical sites, and environmentally sensitive areas with wildlife sanctuaries coming immediately to mind.

More recently, FRA working in concert with the Pipeline and Hazardous Materials Safety Administration (PHMSA) developed the Rail Corridor Risk Management System (RCRMS) whereby the railroads are required to assess the relative risk of their routes for hazmat-bearing trains. From a technical standpoint, RCRMS is a 27-variable algorithm that includes the aforementioned venues to avoid, but it also considers the track class, type, and maintenance; availability to incident response resources; and existing other uses for particular train routings (49 CFR 172). These variables are shown in Table 7.1, below.

For ease of comprehension, these 27 factors can be aggregated into seven general factors as shown in Table 7.2 (Gordon and Young 2016).

Table 7.1 RCRMS risk factor assessment summary

No.	*Description*
1	Hazmat volume transported
2	Rail traffic density
3	Overall route length
4	Railroad facilities
5	Track class, type, and maintenance levels
6	Track geometry
7	Signaling and train control systems
8	Wayside defect detectors
9	Grade crossings
10	Single versus multiple tracks
11	Turnouts
12	Proximity to iconic targets
13	Environmentally significant areas
14	Population density
15	Venues along route(s)
16	Emergency response capabilities and available resources
17	High consequence areas
18	Tracks shared with passenger trains
19	Operating speeds
20	Railroad maintenance and repair facilities
21	Known threats
22	Safety and security risk mitigation measures
23	Alternate route availability
24	Past incidents
25	Overall transit time
26	Training and skills of railroad crews
27	Rail network traffic and congestion

Data source: Federal Railroad Administration

Table 7.2 Categories of rerouting factors

Category	*Description*
Rail shipments	Quantities and properties of hazmat
Train operations	Maintenance, speed, and transit times
Route configuration	Points of conflict and encumbrances
Technology safety measures	Technology-based controls
Rail system network	Potential for conflict and delays
Safety and incident response	Availability of adequate response resources and their response times
Threats, risks, and vulnerabilities	Threat and exposure potential

Source: the authors

Given that the RCRMS seeks to compare the characteristics of pairs of alternatives, the ability to achieve optimality is an elusive objective. Moreover, the best route may well be one that uses, at least in part, the lines of a competing railroad. That eventuality creates yet another tradeoff that needs to be considered, namely that interchange may be overall a more risky alternative than not undertaking the rerouting in the first place. Consider, for example, the issues involving each of the consolidated factors.

Discussion of factors

Although many of these factors are a function of route length, others are much more elusive, and such issues as population or route capacity must be considered. For example, the number of additional grade crossings on an alternate route will cause more points of conflict between trains and highway vehicles. Stadiums and shopping malls could, at certain times, put thousands of people in the proximity of a hazmat train. This section will discuss quantifying measures for each consolidated factor and its components (the 27 FRA factors in Appendix D of 49 CFR 172) for suitability. Moreover, it is entirely possible that there are no feasible alternative routes.

Rail shipments

The initial condition of analysis would be the number of rail cars transported along a corridor in a train and on a daily basis. Here the 20-rail-car limitation for key train speed limits will be a factor.

Train operations

Track either meets or does not meet the standards stated in the FRA classification. This impacts speed, so a single parameter for train operations could be percent of timetable speed and ultimately the overall transit time. Consideration, however, must be given to excessive grade and curvature and their impact on operations, as track geometry impact on FRA class may not be fully addressed in the track safety standards (49 CFR 213).

Route configuration

Route configuration is a function of the physical components of the primary and alternate routes. There is an argument that it has a linear relationship based on route length and number of components. Switches and turnouts, which are a function of yards, team tracks, and other railroad facilities; single vs. multiple tracks; and other uses of turnouts and switches, such as industry tracks – these are all complicating factors when it comes to considering alternative routes.

Grade crossings, where vehicles cross the tracks, depend on active signals and passive roadway signs to minimize the risk of conflict. There is a complexity level with regard to traffic volumes, number of trains, accident rate, and type of protection as part of route configuration.

Rail system network

Rerouting hazmat-carrying trains is not necessarily a simple matter given that the original routing was likely based on distance. Operationalizing the five network factors means finding the shortest available route that (1) keeps hazmat away from passenger trains and (2) avoids traffic congestion that potentially slows or idles trains.

Rail traffic density is a function of each route, and route length would be the actual miles. Shared track with passenger trains could be based on the number of trains per day of each (freight, hazmat, and passenger) and measured in the same manner as density and route length. By addressing these, one arrives at an assessment of congestion on the routes. Alternate route availability and having no alternate route can be looked at in the three conditions: there is an alternate route available, one that is not economically feasible and no possible alternate route at all (Gordon and Young 2016).

Technology safety measures

The ability to control the location of a train, and to have real-time information concerning the integrity of the critical components of the train, is a prerequisite for the identification of any potential alternate routing. In other words, the railroad needs to know where the train is and what the condition of the locomotives and railcars happen to be.

Safety and incident response

This issue is divided into two categories: regular surveillance and crew training (which are either present or not) and the time required for responders to appear at an incident scene. Surveillance and training need to be prerequisites. Response capabilities, specifically defined as the ability to handle hazmat incidents, are measured as the time required for responders to get to the incident and whether the respective capacity is sufficient to address the threats posed by the incident (number and type: hazmat, EMS, fire department, etc.). Maintenance and repair facilities focus on emergency repairs and mobile crews and equipment to handle a hazmat incident and, in particular, heavy rail cars.

The safety and security risk mitigation measures are expressed as the frequency of inspection of the line including on-track and remote monitoring and response time for police and other responders to any point. The level of experience and expertise of the emergency responders, equipment and miscellaneous resources to handle a hazmat incident are measures of adequacy. The skills of the railroad crews need to reflect current and ongoing training.

Threats, risks, and vulnerabilities

This factor looks at the threats and vulnerabilities associated with each, but also as a combination and their relationship to overall risk. For example, the proximity of a hazmat train to venues of iconic and other value and use, and populated high-consequence areas identifies the population and *functions* that could be at risk. Known threats and past incidents address risk and mitigation measures that have been or need to be taken. Environmentally sensitive areas could reflect areas of concern but may not pose a significant risk to the wellbeing of the population within a given distance from a hazmat route.

Concluding observations

While the problem of transporting multiple tank cars of hazardous materials is not new, both the length of the trains and the number of those trains increased dramatically in the period from 2010 to 2015. As mentioned in the previous chapter, the safety record of the railroads is enviable with 99.997 percent of their loads being transported without any form of material release (AAR 2015). However, the concentration of crude oil and ethanol volumes have presented a problem that has required changes to public policies that focus on four distinct approaches: (1) improving the specifications for tank cars whereby they are less likely to either rupture or release their contents in the event of a derailment; (2) identifying trains of particular interest because of the materials that they are transporting, the quantities involved, and the specifications of the tank cars in their consists; (3) providing for the sharing of information about the routing of trains, their frequency, and volume with the respective first responders located along any route; and (4) conducting a risk assessment of the routes being taken by hazmat trains of interest. With respect to routing, the risk assessment will endeavor to use a multivariant approach with the objective function being to find that route where risk is minimized.

Perhaps the most difficult aspect addressed in this chapter is not the cost of investing in new and improved tank cars, nor is it the difficulty of collecting the relevant data needed to conduct meaningful risk analyses. Rather, it is the public policy clash between safety and security whereby the information concerning the training of carrying hazmat would be of interest to not only first responders but those parties wishing to employ such materials in a harmful manner.

The topic addressed in this chapter was originally intended only as a sidebar discussion in the previous chapter, but with the event of the Lac-Mégantic disaster the public discussion became so extensive that a separate chapter became warranted.

Oil spill in Oregon

The Columbia River Gorge is one of the most beautiful spots in North America. An 80-mile canyon where the Columbia River cuts through the Cascade Mountains, the gorge features dozens of waterfalls, high hills with commanding views of the broad blue river, and refreshing fresh breezes on which eagles and ospreys soar. The gorge has been designated a national scenic area: the Columbia Gorge National Scenic Area, managed by the Columbia River Gorge Commission and the United States Forest Service. It is a very special place, unlike any other river canyon or gorge in the nation.

The gorge is also a major transportation artery, with busy railroads on both the Oregon and the Washington sides of the Columbia, often close to the banks of the river. The Union Pacific railroad hugs the south bank in Oregon, while the BNSF's former Spokane, Portland & Seattle line follows the more rugged north side of the river in Washington. Following the course of the river through the gorge eliminates the need for the long, steep grades that otherwise are needed to cross the Cascades.

In recent years, oil has become a significant commodity for both the UP and the BNSF in the gorge. Vancouver, Washington, on the north side of the river near where it makes a bend to the north before emptying into the Pacific, is a major port for transloading oil from rail to marine transport. Not without controversy, plans have been offered to increase the size of Vancouver's oil terminal, to accommodate around 1 million barrels of oil per day – or around four fully loaded oil trains. Controversy has arisen over the potential for environmental hazards, as well as the land use impact the largest terminal, the Tesoro Savage facility, would have on the city of Vancouver.

On June 3, 2016, a 96-car Union Pacific oil train derailed at Mosier, Oregon, in the gorge. The train was moving from Eastport (Idaho) to Tacoma (Washington) with oil from New Town (North Dakota), a town in the Bakken oil field. Fourteen cars derailed and caught on fire, which burned for over 14 hours. About 42,000 gallons of crude oil were discharged. There were some fortunate elements in the crash: it occurred at a spot on the line not directly aside the river, and not in close proximity to a major population center (although a nearby trailer park had to be evacuated). However, the rural location made it difficult for adequate firefighting capacity to respond and bring the fire under control.

An examination by the FRA determined that the cause of the crash was broken bolts in the rail. The FRA was critical of the maintenance of the track and particularly of the braking equipment on the oil cars. Loaded oil tank cars are especially heavy and cause great strain on the tracks. The railroad had run mechanized stress tests on the line every 18 months, with the most recent test in the week before the wreck.

Before the end of June, oil trains were once again moving through the gorge over the UP. As the FRA noted, it had no jurisdiction to determine where rail cargoes could be routed. And the BNSF moved a steady stream of oil trains over the even more hazardous line on the north bank of the Columbia, a route that crosses an even more rugged landscape and in several spots crosses over causeways along the river.

The spill and fire ignited more than oil. It brought into sharp focus the debate over the safety and security of moving oil by rail, and the potential for environmental as well as human disaster. The governor of Oregon has called for a complete halt to the movement of oil by rail until a review of safety concerns has been conducted. Tribal groups concerned with salmon runs and other groups have joined a coalition to fight the movement of oil by rail through the scenic Columbia River Valley. Although the total amount of oil moving by rail is receding, the prospect of major oil terminals along the Columbia and in other locations in the Northwest will keep the controversy alive for years to come.

Notes

1 The term has been used extensively by both print and broadcast media across the country wherever crude trains may run.
2 Industry in this case refers to that combination of the railroads, the shippers, and various leasing companies involved.
3 In 2014 the MMA's assets were sold and a new railroad, the Central Maine and Quebec Railway, was established
4 Canadian Transportation Safety Board, *Lac-Mégantic runaway train and derailment investigation summary*.
5 Note that Canadian and U.S. regulations are substantially similar across many situations, but especially when it comes to transportation equipment standards.
6 RVP is a common measure of the volatility of liquids and is the vapor pressure exhibited at 100° F per ASTM – D-323.
7 Multiple cars of similar materials tendered by a single shipper en route to either a single consignee or perhaps a rail yard for reclassification are referred to as a "cut of cars".
8 Entire trains moving a single commodity are referred to as "unit trains", which most often are hauling coal, coke, or intermodal containers and/or trailers. Unit trains of oil are a relatively new phenomenon.
9 Denotes a design where the tank vessel is an integral part of the car structure rather than being an addition to it.
10 The original knuckle coupler was a boon to safety because of its integrity in two geometric planes. The non-overriding double shelf coupler adds the third plane.
11 Note that because of the amount of cross-border traffic, Transport Canada and the U.S. Federal Railroad Administration collaborate extensively on equipment standards.
12 While ethanol is designated a flammable, it has an RVP of only 3–4, which is significantly lower than crude no matter what the source.
13 This can also be seen as an incentive for shippers and railroads to migrate to the new DOT 117 standard as soon as possible.
14 Acids fall under the classification of corrosive and require equipment different than the DOT 111 specified cars.

Bibliography

Association of American Railroads. (2015). *Railroads 101*. Retrieved February 27, 2016 from www.aar.org/pages/railroad-101.aspx

Bush, G. (2003). *The national strategy for the physical protection of critical infrastructure and key assets*. Darby, PA: Diane Publishing.

Canadian Transport Safety Board. (2014). *Lac megantic runaway train and derailment investigation summary*. Retrieved November 7, 2015 from www.tsb.gc.ca/eng/rapports-reports/rail/rl13d0054/r13d0054-r-3w.asp

Citizens Acting for Rail Safety (CARS) (2016). *DOT 111 tank cars*. Retrieved February 2, 2016 from http://saferails.org/?page _id=51

Gordon, G., and Young, R. (2015, March). *Hazmat routing: Safety and security challenges*. Proceedings of the 2015 Joint Rail Conference. San Jose, CA: American Society of Mechanical Engineers, San Jose State University.

Gordon, G., and Young, R. (2016, April). *Hazmat routing: Alternate routing risk analysis*. Proceedings of the 2016 Joint Rail Conference. American Society of Mechanical Engineers, University of South Carolina, Columbia, SC.

Park, G. (2014, March 16). *Exactly how volatile? Petroleum News Bakken*. Retrieved February 2, 2016 from www.petroleumnewsbakken.com/pntruncate/240780181.shtml

Plant, J., and Young, R. (2015). Railroad infrastructure: Protecting an increasingly vulnerable asset. In S. Hakim, G. Albert, and Y. Shiftan (Eds.), *Securing transportation systems*, pp. 177–188. Hoboken, NJ: Wiley.

Feldman, V. N. (2015, May 8). *PHMSA issues final rule on enhanced tank car standards and rail operational controls for flammable liquid transport*. Retrieved June 4, 2016 from www.vnf.com/phmsa-issues-final-rule-on-enhanced-tank-car

Vantuono, W. C. (2014, February 20). BNSF taking bids on 5,000 oil tank cars. *Railway Age*. Retrieved June 4, 2016 from www.railwayage.com/index.php/mechanical/freight-cars/bnsf-taking-bids-on-5000-oil-tank-cars.html?channel=59

8 Commuter passenger rail

Philadelphia's bustling 30th Street Station sees some passengers come and go trailed by redcaps pushing carts piled high with luggage; however, the vast majority of those using the second-busiest terminal of the Amtrak system will be typically found carrying a purse and/or a briefcase. For nearly two centuries, the majority of U.S. train travel has consisted on trips of less than 100 miles. While intercity travel has gotten much of the press coverage and has traditionally been the romantic stuff that movies are made of, it has been the commuter segment of the industry that has seen the bulk of the ridership.

Passenger rail services in the U.S. are quite unlike that found in most other parts of the world because of the myriad systems operated nationwide. While there are only two systems that are considered true intercity systems, namely Amtrak and the Alaska Railroad, there are many commuter systems. Some of these have relatively few route miles, while others are significantly longer, stretching several hundred miles like multiple spokes connected to a central urban hub – but they also can pass through multiple political jurisdictions each with their own police and other first-responder capabilities. However, the distinction does get blurred given that Amtrak carries a significant number of commuters daily along the Northeast Corridor as well as the Commonwealth of Pennsylvania subsidized Keystone Service stretching from Philadelphia to Harrisburg and California's Caltrain operations. Some commuter lines actually provide intercity riders, by definition, between origins and destinations that may also be interstate – NJ Transit Rail to Port Jervis, NY, as well as Metro-North from Manhattan to points in Connecticut.

While Amtrak is perhaps the best-known U.S. rail passenger carrier, the commuter lines remain of key interest because they have most often been the targets of terrorist attacks, worldwide. This chapter will provide a detailed description of U.S. commuter rail, how it is currently developing, how it approaches the issue of protection, and some thoughts about how it will further evolve. Only then, will the following chapter (Chapter 9) delve into the topic of intercity passenger rail services.

When tragedy visits commuter lines

The most famous disaster associated with commuter railroads is the orchestrated simultaneous attacks on multiple trains in Madrid during rush hour on March 11,

2004. The bombings, later linked to Al Qaeda, killed 191 and injured more than 1,800. The noteworthy variables are density of ridership and time of day. Although this book is primarily about heavy rail operations, thereby excluding light rail lines such as subways and street railways or trams, there have been a considerable number of attacks during the past 20 years that underscore the vulnerability of commuter trains in general, as shown in Table 8.1.

So what makes commuter trains such viable targets? In parsing the nature of operations into the subtopics of (1) users of the system, (2) timing of operations, (3) nature of terminals, and (4) nature of trains, there are reasons why these are preferred targets.

Commuters are typically those passengers that use the rail system on a daily basis consistently between the same pair of origin and destination points. Moreover, they are most apt to use the system *twice* each day, meaning that the most worthwhile timing for targets is likely to be during morning and evening rush hours. Moreover, in modern polyglot societies, commuters are drawn from all socio-economic strata, representing a broad range of ethnicities, meaning that those intent on inflicting harm on the system are not readily standing out. Passengers crowd station areas as well as trains, suggesting that maximum effect can be achieved. In the short-term, an incident shutting down one or more lines has an adverse impact on economic activity, but people still need to get to and from work. When the rail system becomes suspect, riders will turn to other means, namely highway traffic and overcrowded busses.

Table 8.1 Most-significant attacks on commuter rail

Date	*Location*	*Attack method*	*Persons killed*	*Persons injured*	*Responsibility*
July 1993	Baku, Azerbaijan	Bombing	13	42	Armenian separatists
March 1995	Tokyo, Japan	Coordinated attack at five stations	15	50	Cult group
July 1995	Paris, France	Multiple bombings of rail and subway stations over several days	8	200	Islamists
February 2003	Seoul, South Korea	Arson attack in subway station	200	147	Deranged individual
February 2004	Moscow, Russia	Suicide bombing of station	41	120	Chechen terrorists
March 2004	Madrid, Spain	Coordinated bombings of commuter trains	191	1,800	Islamists
July 2005	London, England	Coordinated suicide bombings on trains	52	700	Islamists
March 2010	Moscow, Russia	Suicide bombing of stations	37	65	Islamists

Data sources: www.livescience.com and www.CNNLibrary.com

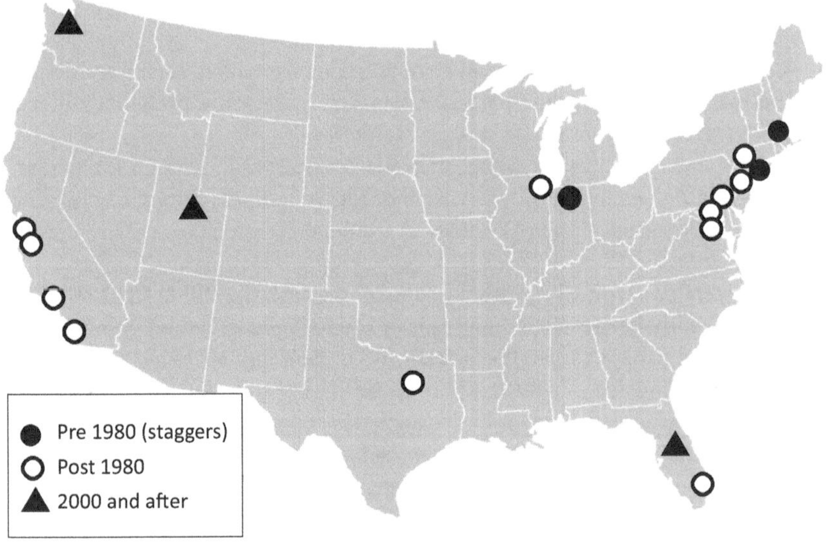

Figure 8.1 Major commuter rail lines

Source: compiled by the authors

In the last 30 years, with the growth of suburbs that now often extend 100 miles or more from center city to their outer limits, commuter rail operations can still be intrastate, but increasingly they are interstate in nature. In the U.S., examples abound with New York's Metro-North extending well into Connecticut; NJ Transit operating in both Pennsylvania and New York; and Chicago's Metra going into both Indiana and Wisconsin. The number of commuter rail operations in the U.S. has been increasing in recent years. Those well-established systems of Boston, New York–New Jersey, Philadelphia, Baltimore-Washington, Chicago, Los Angeles, and San Francisco are descendants of the freight and passenger railroads that prevailed up through deregulation in 1980 and later spun off to regional operating authorities. These are recently being joined by nascent yet expanding operations in Seattle-Tacoma, Central Florida, Albuquerque, Dallas / Fort Worth, and Salt Lake City. Figure 8.1 shows the locations and approximate chronology of these systems.

Nature of operations

Economists would ascertain that commuter rail operations have lumpy demand, meaning that they are busiest between 6:00 and 9:00 a.m. and between 4:00 and 7:00 p.m., with the other hours of the day being relatively slack. Commuter stations run the gamut between the large and opulent, such as Philadelphia's 30th Street with its grand high ceilings, art deco fixtures, and large murals to those with outright

Spartan facilities consisting of nothing more than an open, uncovered platform that may or may not be at train level. In between these two extremes are those stops with ancient wood frame stations that may be shuttered or the more modern paved platform with bus-stop-like shelters offering minimal protection from the elements. Most will have nearby parking that nowadays is often filled to beyond capacity.

The typical commuter is often someone working in an urban area in some middle management capacity such as finance, accounting, or marketing. They make the daily commute in order to flee the high cost of living of the cities and to achieve a lifestyle that enjoys a modicum of safety and where their children can have access to better public schools. The tradeoff is the wear and tear of the daily journeys, the time spent, and the cost of the transportation.

Line density

Commuter rail lines are expensive to operate. The original ones that were part of freight railroads were for the most part money losers. In the middle of the last century, the freight railroads sought to exit commuter transportation with light density lines being the first to be identified for abandonment. The corollary here is that rail is preferred for the densest routes, while other modes, such as bus or light rail transit, are the choices for lighter ones and more close-in commuters. Heavier densities mean greater ease of movement by perpetrators as well as maximum impact resulting from their actions.

Economic impact and the fear factor

When commuter rail lines fall victim to natural disasters, as was the case in 2012 with New Jersey and Hurricane Sandy or potentially anywhere with a terrorist attack, the system may, depending on the severity of the event, be out of service for some period of time unless there is some level of system redundancy. The loss of commuter transport capacity can have a twofold impact: reducing or outright cutting off a workforce necessary to sustain economic activity, and the loss of customers to those businesses that support the daily needs of those workers.

To the commuter rail line it means a loss of revenue all the while that the costs of maintaining the fixed assets continue. Variable costs will not be incurred, including operating labor, power for electrified lines, and diesel fuel, all of which represent secondary economic impacts.

Natural disasters – such as earthquakes, hurricanes, and other weather-related events – may have a greater regional impact depending on their relative severity. Terrorist attacks, while more focused with respect to targets and resulting damage incurred, could be expected to have greater economic impact because of the psychological effect, meaning that commuters will likely seek other means of transportation because of the associated fear factor. In addition, an attack on just one train or in one station will often result in the entire system being shut down and carefully checked for additional evidence of tampering.

While the ridership will eventually return to pre-incident levels, this could take time and is a phenomenon not unlike the drop in airline passenger volume following the events of 9/11. The economic impact is a direct result of the fear factor, which is especially compelling in societies where personal safety is generally accepted as a given. The presence of pickpockets and other perpetrators of petty crimes is deemed more of an inconvenience when compared with the potential for injury or loss of life on a larger scale. Damage to infrastructure may or may not be a factor when one compares the circumstances of the Tokyo sarin attacks with the Mumbai or Madrid bombings. Nevertheless, the omnipresent fear factor is an objective.

Newsworthiness

Catastrophic events, natural or otherwise, always head the list for major news coverage because of sensationalism. Images of newscasters interviewing homeowners surveying flooded neighborhoods makes for prime-time television news coverage. Terrorist attacks on transportation infrastructure also get media coverage, but coverage that is in fact one of the intended objectives of staging the incident: generating fear in the minds of the populace and providing publicity for the political cause of those committing the act.

While natural disasters occur and society is left cleaning up the aftermath often for some period of time, terrorist strikes continue their intended impact of gaining publicity, which could provide value lasting years afterward. That publicity also serves to feed the fear factor as well as prompt society to establish countermeasures. The events of 9/11 not only caused the establishment of the Transportation Security Administration to screen primarily airline passengers, but the formation of a whole new federal department, Homeland Security, that came with a realignment of responsibilities and regulations for myriad other units including the U.S. Coast Guard, Customs, Federal Emergency Management Agency, and Border Patrol.

The difference between the types of threats and their reactions are found in Figure 8.2, below; note that human-caused events, whether industrial accidents or acts of war, result in a government reaction that can generally be termed *regulation.*

Impact and the ability to protect

Protecting railroad infrastructure is not a simple task, because of the magnitude of its geographic coverage, the necessity of keeping it accessible to its passengers, the absolute number of nodes (stations, terminals, and junction points), and the longevity of many of its structures. Much of the right-of-way and key infrastructure components were originally built in either the heyday of the railroad expansion in the mid to late nineteenth century or in the period of the nation's rapid economic expansion during the first third of the twentieth century.

Routes are found where they are for two reasons: proximity to population, meaning passengers in urban areas and ease of building through rural terrain, which

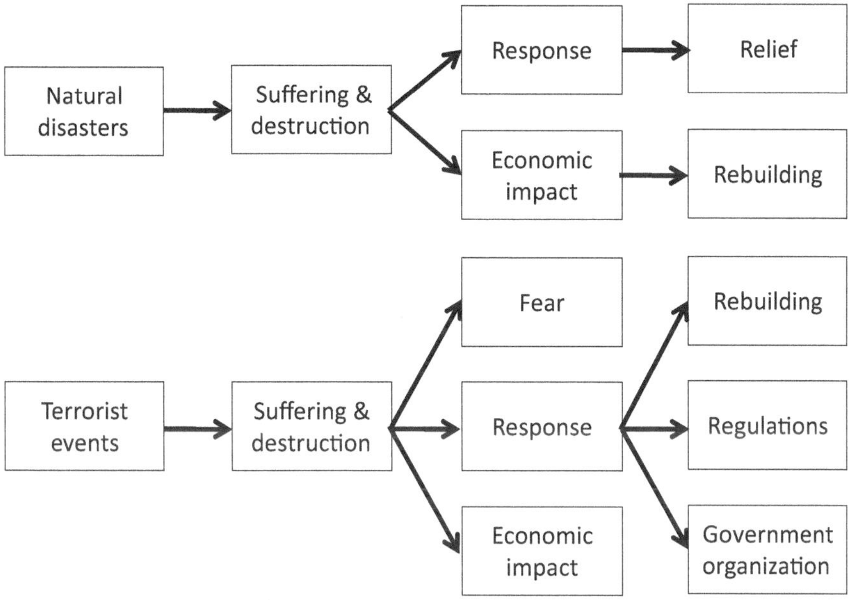

Figure 8.2 Disaster responses

Source: compiled by the authors

usually meant level terrain that minimized the amount of gradient required, the gentlest grades for trains to climb and descend, and, if possible, the fewest number of tunnels and bridges. These reasons, however, give rise to vulnerability from manmade and natural-caused sources. Of these, particular attention needs to be paid to flooding because rail routes typically follow rivers because of the gradient.

Relocating rights-of-way is difficult, if not impossible, in urban areas, but any relocation cannot be removed from the proximity of passengers. Hence, the conundrum is keeping the infrastructure out of harm's way while having the railroad serve its intended need.

Geographic coverage

As the U.S. industry built railroads during the last two centuries, it became the major source of improved transportation given that air and road had not yet evolved as modes. Communities vied for railroad connections because they knew the economic activity that would follow. Eventually, the network was not only built but overbuilt. Mergers and acquisitions began, shortlines and regional railroads disappeared, and large systems emerged. Many of those systems operated three different segments of the industry: freight, intercity passenger, and commuter lines. While sharing some equipment, infrastructure, personnel, and marketing skills, each segment also had sizeable differences, but especially so with economic regulation. Commuter lines were often the domain of the states, whereas much of the rest was under the control of the sunsetted Interstate Commerce Commission (ICC) – now to some extent the Surface Transportation Board.

This is germane because when the railroads were economically deregulated, the process occurred in three major stages. In 1970, Congress passed the Rail Passenger Services Act that established the National Rail Passenger Corporation, better known as Amtrak, to operate the money-losing intercity trains. During the ensuing decade, the railroads, still under the economic control of the ICC, continued to suffer financially. The Staggers Act became law in 1980, and the railroads were able to shed excess infrastructure and sell off or abandon unprofitable lines; however, they were still operating commuter trains, albeit with state subsidies in many metropolitan areas such as Philadelphia, New York, and Boston, but nevertheless remained a financial burden. In 1981, Congress passed the Northeast Rail Service Act that transferred commuter operations from Conrail, the previously established quasi-government corporation, to newly established state and regional quasi-governmental agencies. These agencies acquired rolling stock, stations, maintenance facilities, personnel, and some, but not all, of the rights-of-way. At about that time, other state legislation established myriad commuter agencies independent of their former freight railroad parents in other parts of the U.S.

All of the preceding history is necessary to gain an understanding of why the commuter lines are located where they are and why they often share infrastructure with both Amtrak and the freight railroads. Note that Table 8.2 has most of the nation's commuter railroads being established during the 20-year period after the Staggers Act. Table 8.2 traces that lineage, but readers should also be aware that Amtrak also carries a significant number of daily commuters even though the topic of intercity passenger rail is covered in the following chapter.

Multiple use rights-of-way

Perhaps one of the most significant contributors to the complexity of protecting commuter lines is that their rights-of-way are most likely shared with others. Such sharing occurs as either other rail services operating over the commuter line or perhaps the commuter line as a tenant on either Amtrak or a freight railroad. Added to the equation is that a commuter line may have a dedicated line but that there are others that share the right-of-way even if they do not use the commuter line's rails per se.

In locales such as the Northeast Corridor, commuter lines SEPTA, MARC, and NJ Transit operate over Amtrak; however, where they own their own lines, there may be freight operators such as Norfolk Southern, CSX, or a shortline such as Morristown and Erie serving local industries. There are stretches of rail where the freight railroads and the commuter lines have their own rails but are running side-by-side for some distance. Perhaps an anomaly is the regional (not exactly commuter) line that is planned for connecting Orlando with Miami, which will be operated by Florida East Coast, where one could expect both freight and commuter trains to be sharing the same tracks.

Table 8.2 Genesis of commuter rail operations post-Staggers

Name	*Year founded*	*States served*	*Route miles*	*Predecessor railroad(s) or track owner*	*Operated by*
Metro-North (MNCR)	1983	CT, NY	384	Conrail (New Haven, New York Central)	Quasi-gov't authority
Metropolitan Transportation Authority–Long Island Railroad	1836	NY	594	Long Island Railroad (From 1900 owned by Pennsylvania Railroad and successors)	Quasi-gov't authority
New Jersey Transit Rail	1983	NJ, NY, PA	544	Conrail (Central of New Jersey, Erie Lackawanna, Penn Central, New York Central, Pennsylvania-Reading Seashore Line)	State gov't
Southeastern PA Transit Authority Regional Rail (SEPTA)	1983	NJ, PA	280	Conrail (Penn Central, Reading)	Quasi-gov't authority
Metropolitan Rail Corporation (METRA)	1984	IN, IL, WS	487	Chicago Northwestern, Union Pacific, Burlington Northern, Rock Island	Quasi-gov't authority
Peninsula Corridor Joint Powers Board (Caltrain)	1987	CA	51	Southern Pacific	Quasi-gov't authority
Southern California Regional Rail Authority (Metrolink)	1992	CA	388	Southern Pacific	Quasi-gov't authority
Maryland Dept. of Transportation (MARC)	1984	DC, MD	202	Conrail (Penn Central), Chessie	State gov't
Virginia Railway Express	1992	DC, VA	97	Chessie, RF&P	Franchisee corporation
Tri-County Commuter Rail Authority (Tri-Rail)	1987	FL	72	CSX, Florida East Coast	Quasi-gov't authority
Trinity Railway Express	1996	TX	34	BNSF	Quasi-gov't authority
Central Puget Sound Regional Transit	1995	WA	80	Southern Pacific	State gov't
North San Diego County Transit Dev. Board (Coaster and Sprinter)	1976	CA	63	Southern Pacific	Regional gov't
New Mexico Railrunner Express	2006	NM	97	BNSF	State gov't
Massachusetts Bay Transportation Authority (the T)	1974	MA	388	Boston and Maine, New Haven	Quasi-gov't authority
Music City Star East Corridor Commuter Rail	2006	TN	32	Nashville and Eastern	Regional COG

(*Continued*)

Table 8.2 (Continued)

Name	*Year founded*	*States served*	*Route miles*	*Predecessor railroad(s) or track owner*	*Operated by*
South Shore Line	1989	IL, IN	90	Chicago South Shore and South Bend Railroad	Quasi-gov't authority
Northstar	2009	MN	40	BNSF	Quasi-gov't authority
Capital MetroRail	2010	TX	32	Southern Pacific	Quasi-gov't authority
UTA Frontrunner	2012	UT	88	Union Pacific	State authority
Westside Express	2009	OR	15	BNSF, Oregon Electric	Quasi-gov't authority
DCTA-A Train	2011	TX	21		Quasi-gov't authority
SunRail	2014	FL	32 (still being built)	CSX	Quasi-gov't authority
Altamont Corridor Express	1997	CA	85	Union Pacific	Regional COG

Data sources: Federal Transit Administration, *Trains Magazine*, various operator websites

Rail lines are useful corridors for others. It is not uncommon for light or heavy rail transit services to have parallel lines running through the same right-of-way as is found in some part of the New York metropolitan area with Metro-North, the Long Island Railroad, and the New York City Subway System or the Massachusetts Bay Transportation Authority commuter rail and Red Line heavy rail transit. In northern New Jersey, the Port Authority Trans Hudson (PATH) occupies some common right-of-way with NJ Transit just as the San Francisco's Caltrain shares stretches with both Union Pacific and Bay Area Rapid Transit (BART).

Finally, there is the utility corridor aspect where electric, fiber optic telephone, gas, and water and sewer share the rail right of way – either longitudinally or transverse. The rail lines are convenient for these purposes because the routes are largely unobstructed with few conflicting underground or aerial structures that need to be accommodated. This sharing is not free, because the railroads collect rent from the utilities, but it is mentioned here because the vulnerability of the rail line also becomes such for the respective utility operators.

Shared rights-of-way, whether as multiple types of rail operations using common track, other railroads including rail transit using adjacent tracks, or utilities occupying underground or aerial routes for transmission lines, the density of use increases the attractiveness of such infrastructure to manmade threats as well as the vulnerability to natural ones.

Infrastructure components

All railroads require bridges, tunnels, and culverts to accommodate the geography of their respective operating area. All railroads need these to overcome naturally occurring obstacles such as rivers and mountains. Commuter railroads are confronted by significantly higher incidents of manmade disasters in the form of roadways, commercial and residential structures, and other railroad rights-of-way and its inherent infrastructure.

Infrastructure components are more expensive to maintain; take a greater investment in terms of both time and investment to construct, maintain, or replace; and pose a greater vulnerability to both natural and human-caused threats. Moreover, an incident that removes a mile of freight line in relatively open terrain can be overcome with the rerouting of trains on other lines even if these are owned or controlled by other, often competing, railroads. The loss of a bridge on a key commuter line may have no such alternative available, and it is, therefore, not uncommon to hear that commuters are diverted to one or more bus lines. Some infrastructure represents a significant bottleneck when it is used by more than one railroad, such as the tunnels linking New Jersey with New York City's Penn Station. Originally built by the Pennsylvania Railroad in the early part of the twentieth century, utilization of the tunnels is at capacity, there are no practical alternative rail routings, and an event as simple as a transformer failure can lead to major disruption of commuter operations.

Terminals and stations, as previously established, may be major structures accommodating numerous trains on a daily basis or are isolated stops with the most minimal facilities. Loss of the former, by whatever means, can pose a major disruption of service given that the structure also has the potential of removing one or more tracks from service. Moreover, a major station will take far longer and at a much greater investment to repair or replace. Being more complex, it will be more difficult to protect as well, even though the barebones distant suburban station will have less protection and consequently be a more attractive location for terrorist access.

Maintenance

A typical commuter railroad is likely to have maintenance facilities that will be used for both forms of rolling stock: cars and locomotives. Whether a system uses both electrified equipment and diesel power is probably immaterial as illustrated by NJ Transit's mammoth Meadowlands Maintenance Complex (MMC) built in the mid-1980s. The single facility will better achieve economies of scale, but on the other hand it becomes a de facto bottleneck should it be hit with a natural disaster or a terrorist act. The former eventuality was clearly the case for NJ Transit with the MMC, built just feet above the water table, when the storm surge from Hurricane Sandy hit in 2012. Not only was the equipment being stored there inundated, but the needed capacity to repair other equipment that had been affected from around the system was also lost. This may be in stark contrast with the Class I freight railroads that will have multiple maintenance facilities, often specializing in rail car or locomotive repair.

Rail yards for commuter railroads are normally established at the ends of lines and near the key maintenance facilities. Loss of these facilities may mean the inability to layover trains and to stage them for the next commuter surge. Whereas freight lines use yards to not only stage trains but make up various consists, much of this activity is a much more seldom occurrence. Still, the loss of rail yards more due to natural disasters can be a major event for the commuter railroads. Again, referring to Hurricane Sandy, the storm surge affected the yards adjacent to the MMC where tracks were mere feet about the water table.

While an asset that is likely invisible to the casual observer, communications and signaling (C&S) is a component of commuter railroads that represent a two-fold value: they facilitate safe operation of the line, and they permit trains to operate in closer proximity, thereby enabling more efficient use of the existing capacity. Commuter railroads, because of their generally limited number of track miles (see Table 8.2), operate with a single main dispatching center[1] and employ computer systems to align switches and set signal aspects governing train movements.[2] Loss of the system stemming from a natural disaster or a terrorist action will essentially shut down the line. Power outages may not pose the same short-term threat because backup generating capacity is often provided for.

The key word that describes the justification for much of the infrastructure is *density*. It is a temporal density of operation meaning that there is a bimodal distribution for the frequency of service as well as a likely relative size of train when demand is at its highest.

Commuter line ownership

Previously, the commuter lines were an integral part of railroads, expressed as a general term. Since deregulation they have been run as units of state government as in the case of MARC and Caltrain; as a separate authority like NJ Transit, Metro-North, and the MBTA; or as a hybrid where the state owns the right-of-way but the operation is contracted out to a private firm. There have been those situations in the past where the commuter line is privately operated, but with subsidies from state government. As noted previously, the planned line from Miami to Orlando in Florida will be owned and operated by a private firm, the Florida East Coast Railway. Two new operations still proving their economic viability are the New Mexico Railrunner and the Music City Star East Corridor in Nashville. Projects still in the planning phase are a commuter line to be operated over the Alaska Railroad, and one from San Antonio to Austin, TX. Also, in the past, a modest commuter operation was provided by the New York, Susquehanna, and Western Railroad for Syracuse, New York, but this is now defunct.

Regulatory structure

At the federal level, commuter railroads are regulated from the safety, security, and financial perspectives. For safety, the Federal Railroad Administration retains jurisdiction with the Federal Transit Administration having responsibility for reviewing all aspects of financial operations. Of interest here is the Transportation Security Administration, which under the Homeland Security Act of 2002 is charged with assuring the security of the traveling public.

Up until September 11, 2001, the FRA was the primary agency for rail safety with its jurisdiction stipulated by the Federal Railroad Safety Act of 1970. At that time security was defined as a subset of safety. In 1994, Congress passed the Federal Railroad Safety Authorization Act of 1994 (49 U.S.C. 20151) with the intention of addressing the growing problem of persons trespassing on railroad property. The Homeland Security Act of 2002 established the Department of Homeland Security with the Transportation Security Administration transfered from DOT. TSA, empowered to deal with security for all modes of transportation, was also paired with the Office of Domestic Preparedness charged with making risk assessments and providing funding to state and municipal governments under the Urban Area Security Initiative (GAO 2006). Of the nearly $18 billion awarded to date, less than $500 million has gone to transit systems under the Transit Security Grant Program. One needs to conclude that given the number of systems cataloged earlier in this chapter and recognizing their geographic scope and daily ridership, the federal funding in light of the attacks on systems in other parts of the world appears scant at best.

While not all of the federal effort, including that of TSA, is focused on air travel, the fact remains that most of the budget expenditures are indeed devoted to that. One argument is that the air transport sector may have higher visibility and that any terrorist act bringing down an airliner has a much higher news value and is thus of much more concern to both the traveling public as well as the public at large. As noted in *Detour Ahead: Critical Vulnerabilities in America's Rail and Mass Transit Security Programs*, there needs to be an ongoing robust dialogue between government and other interested parties in the rail security network.

Policing efforts

As has been noted elsewhere, railroads, whether private corporations or quasi-public transportation authorities, maintain their own police forces. Commuter railroads clearly fit into this category and often have a unique combination of advantages and disadvantages to their respective environments. Many commuter railroads have limited geographic coverage, meaning that their activities are limited to a single state, but most often to a single metropolitan area – although note that some commuter railroads cover extensive geography, with NJ Transit and Metro-North coming immediately to mind. The trains operate the same schedules with the similar consists for five days each week. The task of their railroad police is to protect railroad property, the passengers, and their possessions, plus there may still be targeted freight on adjacent tracks and passing sidings.

Where commuter railroad police have a disadvantageous environment is in the disproportionately large number of nodes (stations and junction points) that they operate for their size, the trackage shared with freight railroads and Amtrak, the large number of passengers clustered in morning and evening rush hours, and the typical crowded urban setting for much of the rights-of-way.

With a relative geographic focus, many commuter railroads also have a significantly fewer number of public sector police agencies with which to establish relationships. Contrast this, for example, with the myriad state, county, and municipal police agencies that a Class I freight railroad such as Norfolk Southern and Union Pacific must attempt to collaborate with.

It is a widely accepted notion that commuter railroads will suffer attacks to their trains and in stations, whereas the potential for death and destruction is greater when intercity railroads receive attacks to their track structures. The corollary to this is that terrorists chose between velocity and density. Moreover, mechanical means of sabotage of creating derailments are far more effective than explosive devices (Kaiser 2011). For intercity trains, and this is explored further in its own chapter, railroads employ two means: (1) a sweeper, which is an empty train or hi-rail vehicle that is run through the system to uncover any potential cause of derailment, and (2) many systems now employ a technology that stops trains in the event that a bomb is detonated prior to arrival in a station, a development that forces terrorists to execute more highly coordinated and timed attacks.

Finally, commuter trains employing trainsets with semi-permanent couplers tend to remain more intact than those with knuckle couplers. In the case of the latter, most of the death and injuries stem from crush and impact fatalities occurring when trailing cars telescope into preceding cars.

Conceptual framework for thinking about infrastructure

In September 2011, the Department of Homeland Security published the National Preparedness Goal (NPG) that identified five mission areas, which are also useful for layering those various initiatives necessary for sustaining the nation's railroad infrastructure. Each of these is briefly defined as follows in Table 8.3.

Throughout this book, we have endeavored to paint the problem of protecting railroad infrastructure as not a singular task despite the fact that intercity passenger, commuter, and freight railroads, at least in the U.S., will often share the same rights-of-way and often the same tracks, and safety and security responsibilities.

Commuter railroads are characterized by wide variance in demand during the day with much of their activities clustered during morning and evening rush hours. While passenger trains as targets for terrorist activity can be generally characterized by velocity and density, commuter operations are a matter of the latter as many of their trains and stations are often crowded with travelers. The users of commuter railroads are often regulars riding trains between the same pair of origin and destination stations multiple times each month, many times for years on end.

Commuter railroads are networks serving particular regions and are often operated by multi-state agencies. Policing these networks is particularly challenging due to the expansive area that the railroads serve, but also due to the nature of their operations where stations may be either large bustling terminals or isolated stops barely worth the designation of station given the lack of any structures. With the former, policing is difficult due to the large volume; with the latter, the

Table 8.3 Preparedness goal mission areas

Initiative	*Definition*
Prevention	Reduce the risk from human-caused incidents and minimize the potential for known risks to affect normal business. Risk, in this case, can be parsed into potential targets, physical location, and where lack of security may exist. Risk may be defined as the product of potential multiplied by severity.
Protection	Eliminate, reduce, and mitigate threats or hazards to people, property, and the environment shielding key resources from intentional attack.
Mitigation	Provide a strategy to lessen the effect to an organization of threats and hazards.
Response	Develop a plan that encompasses the human and material resources necessary for confronting threats and hazards once an incident has occurred assuring that injury to people and damage to property is minimized.
Recovery	Establish those relationships with both internal and external entities whereby the organization can resume normal operations at the earliest feasible time.

Data source: U.S. Federal Emergency Management Agency

inactivity – note that the concern is that a terrorist will board at a remote station where there is minimal security and achieve access to a main terminal, much as what occurred in Madrid. Yet it is the day-in and day-out vigilance that for the most part protects these systems. Video surveillance needs to be present in stations and onboard trains.

Vigilant passengers are an asset that needs to be continuously leveraged because it is the frequent user of the system who may well be the first to notice individuals or articles that are not normal. Hence, the "See Something, Say Something" campaign is used widely. Engaging such passengers to motivate them to be extra sets of eyes and ears may prove an ongoing challenge as these individuals may be sufficiently complacent or perhaps just do not want to be involved. Conversely, those individuals who are knowledgeable of train operations and have an ongoing interest are the hobbyists, otherwise known as the railfans, who often may engage in photography or just plain train watching. As discussed elsewhere in this book, the BNSF, Amtrak, and Norfolk Southern have reversed their previous objections to having these individuals observe operations to now openly embracing their participation with registration systems and even identification.

Chapter summary

Protecting commuter lines may well be the most challenging of any of the types of railroads that we have purposely defined as heavy rail systems. There is no single answer possible, but rather there needs to be a constant awareness on the part of everyone coming in contact with the commuter rail systems. We have repeatedly noted that multiple types of railroads use the same rights-of-way, which also suggests that there needs to be a level of communication between these organizations when it comes to protecting the infrastructure. Clearly, the railroad police need to engage in better collaboration with each other as well as with state and local police agencies. Some of this has begun to transpire with fusion and information-sharing centers established at both state, regional, and industry levels.

Finally, we advocate that while there are common strategies available, including those shared by multiple railroads, no single group of approaches can be universally applied across all systems. From a public policy standpoint, protecting the commuter rail systems, much like protecting the entire U.S. rail network needs to become a higher priority with regard to the legal penalties for persons committing vandalism or trespassing. While the air transport sector is vastly different from rail, there are some similarities that suggest that best practices should be observed, adapted, and then adopted.

Passenger rail attacks in India

Railway operations in India have been common targets of terrorists since the partition of 1947 that created the modern nations of Pakistan and India.

The railway system in India is extensive, busy, and vital to the nation's economic and social life. It is also notoriously prone to accidents due to crowding, poor maintenance of infrastructure and equipment, and density of train movements. Moreover, the past decade has seen an escalation of terrorism directed against the Indian railway system, motivated primarily by the lingering Indian-Pakistani hostility, but also by Maoist rebels in eastern India. The worst attacks are described below.

The Western Line of Mumbai's Suburban Railway Network is one of the busiest commuter operations in the world. On July 11, 2006, it was also the scene of one of the best-planned and executed terrorist attacks ever to be conducted against rail passenger operations. During evening rush hour, seven bombs exploded on crowded commuter trains heading back from downtown Mumbai to residential areas to the north of the city, killing 209 individuals and injuring over 700. The bombs were placed in pressure cookers in first-class compartments. Each contained a 2.5 kilogram mix of RDX (cyclotrimethylene-trinitramine explosive) and ammonium nitrate, a lethal combination exacerbated by the pressure created within the pressure cooker. It was not considered a suicide attack, and despite intensive police investigations that pointed to the Islamic terrorist groups Lashkar-e-Taiba and the Students Islamic Movement of India (SIMI) the perpetrators were never apprehended.

Commuter rail operations present an easy and inviting terrorist target, but the symbolism of a train linking India and Pakistan was a probable reason for the attack on the so-called "Peace Express," the Samjhauta Express, on February 19, 2007. The train is a twice-weekly run between the Indian city of Delhi and the Pakistani city of Lahore, with a border crossing at Attari (India) and Wagah (Pakistan). On the date of the attack, the train was estimated to be traveling at 60 miles per hour near the Dirwana station in India when a series of IEDs exploded in the rear cars, killing 68 and injuring many more. Most of the victims were Pakistani civilians, although several Indian security officers were also killed. Suspicion for the attack fell on both Hindu and Muslim groups. A confession was later obtained from a Hindu leader, Swami Aseemanand, but suspicion arose that it was forced. The Pakistani terrorist group Lashkar-e-Taiba was also suspected, but no clear evidence was ever found to implicate them.

It is not just trains that have been attacked in India. In November 2008 a team of terrorists entered Mumbai by boat from Pakistan and launched a series of attacks in the city. A group of two entered Mumbai's historic downtown rail station, the Chhatrapati Shivaji Terminus, formerly known as Victoria Station, and sprayed the crowded terminal with gunfire, killing 58 and injuring 104 in a 90-minute attack. The gunmen then engaged police in a series of gun battles that left one terrorist dead. The second, Mohammed Ajmal Kasab, was captured and later hanged in 2012. Before his death he implicated the Pakistani state intelligence service and Lashkar-e-Taiba in the planning of the event.

While religion remains a primary basis for terrorist events in India, it is not the only source of conflict. In eastern India, particularly in the province of Bihar, Maoist rebels have launched upwards of 100 attacks on trains, rail infrastructure, and railway personnel in the past five years, blowing up tracks and trains in order to disrupt the economy and destabilize politics in the region to advance their rebellion against the Indian government.

Note, however, that while India has had a significant number of attacks on its railways, recent events in Russia, Spain, and even Germany have elevated the value of passenger railroad passengers, trains, and infrastructure as terrorist targets.

Notes

1 Although backup dispatching centers are common.
2 Positive train control, or PTC, is discussed elsewhere in this volume.

Bibliography

CNN Library (2014). *Spain bombings fast facts*. Retrieved from http://cnn.com/2013/11/04/world/europe/spain-train-bombings

Federal Transit Administration. (2006, November). *Commuter rail safety study*. Retrieved from www.fta.gov/printer-friendly/12537+12949.html

Informs (Institute for Operations Research and the Management Sciences) (2012, October 11). *Terrorism risk greatest for subway/rail commuters*. Retrieved January 1, 2015 from www.informs.org/About-INFORMS/News-Room/Press-Releases/Terrorism-Rail-Risk

Kaiser, K. (2011, August 11). *High speed rail security needs a different approach than commuter rail, Mass Transit*. Retrieved January 11, 2015 from www.masstransitmag.com/article/10317151/high-speed-rail-security

Kabak, B. (2007). Subways remain unsecured in commuter rail security beef-up. *Blog*. Retrieved January 11, 2015 from http://secondavenuesagas.com/2007/03/22/subways-remain-unsecured

Office of Security Policy and Industry Engagement, Transportation Security Administration. (2013). *Mass transit and passenger rail security overview*. Retrieved January 1, 2015 from www.tsa.gov/stakeholder/mass-transit-and-passenger-rail

U.S. Government Accountability Office (USGAO) (2006, March 29). Passenger rail security: Evaluating foreign security practices and risk can help guide security efforts (GAO-06-557T). Retrieved December 7, 2006, from Government Accountability Office via http://www.gao.gov/htext/d06557t.html

9 Intercity passenger rail

In the previous chapter, there was an extensive discussion of the commuter rail network, a diverse combination of entities with some being interconnected and others remaining isolated. Some share rights-of-way with either freight railroads or Amtrak, or both. This chapter opens with a scenario for how a terrorist attack on a rail passenger service might unfold. Then, after a discussion of the history of rail passenger service in the U.S., we provide a primer on the public policies that have been put into place to protect trains. Finally, the chapter concludes with a detailed discussion on the functioning of the Transportation Security Administration, which is better known as the TSA that inspects passengers and their luggage about to board aircraft. TSA has a much lower profile with the railroad passenger, but that is probably because there are, among other things, more airline passengers flying daily than intercity train riders and the difficulty in screening at terminals and wayside stations.

A terrorist scenario

The young man waits nervously in the long line at Washington Union Station already queued up for the departure of a regional train for New York and Boston, with stops at New Kensington, BWI, Baltimore Penn Station, Newark Delaware, Wilmington, Philadelphia, Trenton, Newark NJ, and New York's Penn Station. Most of the travelers in line are trying to get out of DC before the commuter rush heats up later in the afternoon. There is still a half hour to go, but the line already extends into the aisle, past the fast food and juice stands. The seating around the gatc is limited and filled with the usual train travelers: students, tourists from a dozen or more countries, older folks still not inclined to airline travel, professional men and women already finished with the day's business at the nation's capital. It's a hot day, and the atmosphere has the heavy feel of a Washington summer.

The young man feels the tension as he waits for the gate to open. He'd filled his backpack carefully before boarding the Metro train at Dupont Circle an hour earlier. His ticket, purchased online with a phony name, is only to Baltimore-Washington International Airport (BWI): no need to waste money when a longer trip is not needed. In the days earlier, since arriving in Washington, he'd made several trips to Union Station to get a feel for the place, the security arrangements, the police dogs and patrols. Nothing like the security at the airport, he concluded. Stay focused, look innocent, and you are not likely to be noticed.

He'd seen the sign that said IDs must be shown, but noticed as well that the crowd just surged forward once the gate for the train departure was opened, ten minutes or so before scheduled departure. How little attention was given once you showed your ticket and scrambled for a good seat. How little notice was given to baggage flung into the overhead compartments or just dropped on a seat.

Now the gates have opened, and the young man walks in a normal manner toward the middle of the train. He steps into the vestibule, moves forward, and drops the backpack under a seat, then keeps walking forward, to the front of the car and out as a few passengers trying to board give him nasty looks as he steps off the car. Not good – they may remember him, note his Mediterranean coloring and facial features. But too late now – the clock is literally ticking.

Right on time, the train departs. The young man watches the last car move slowly, and then the train picks up speed as it heads north. By the time he takes the Metro back to his friends' apartment, the news should be showing the first shots of the explosion near the BWI stop. Then he'll let everyone know over the Internet what he has done to avenge the deaths in Iraq and swear allegiance to ISIS. First, though, a stop at a juice stand for a cold drink.

Unbeknown to him, one of the students heading north to Delaware noticed the backpack after she settled into the seat behind it. She asked if anyone knew if someone was sitting there, and was told that someone had dropped it and had not been seen since. She contacted the conductor as the train got underway, and soon the train crew notified Amtrak police that a suspicious object was on board. An emergency stop a few miles north of Union Station and the backpack was handed over to the bomb squad as police questioned passengers about the individual who dropped the pack off. With a description in hand, given by one of the boarding passengers who had been bumped by the young man as he detrained, police at Union Station are notified, and the young man's decision to wait a few minutes for a drink before leaving the station proves to be an unwise one, as police apprehend him at gunpoint.

A fortunate ending for a possible act of terrorism, but nevertheless a close call. The backpack in question might have held a bomb, or possibly a chemical or biological agent. Or the individual might have been low-tech, preferring to just pack a gun and ammunition and start shooting people on the train. Or one might say the scenario is unlikely because planes, not trains, are the historical targets of terrorists in the U.S. Nonetheless, it is an undeniable fact that passenger trains are a potential target – some would say a ripe one – given the ease with which a determined terrorist can avoid detection and wreak havoc in a variety of ways. Targeting the train is one approach, but we can envision targeting the right-of-way, stations, train crews, signaling devices and systems, freight or light rail trains on adjacent parallel tracks – in short, a plethora of ways that intercity passenger trains can be targets for terrorist attacks.

Background

While attacks on passenger trains in the U.S. have been virtually non-existent, passenger trains internationally have been targets of terrorist activities since the second half of the nineteenth century. In times of peace and war, enemies of the regime in power provided most of the attacks, beginning with Polish nationalists rising up against Czarist Russia in the 1860s. In the century and half since then, little has changed in terms of threats, motives, and methods. What started in the Russian Empire spread to other nations and continents rapidly and then stabilized until the current wave of disorder and terrorism in the past quarter century sparked a new and still-emerging wave of violence against passenger trains.

Why were trains one of the first targets of terrorism? Why do they remain such a common object of terrorists? Is there a common thread to threats to passenger rail around the globe, or are they endemic to certain societies and locations? What can be done to predict, thwart, and recover from terrorism directed at passenger rail? What specifically is the nature of threats to long-distance, intercity passenger rail in North America, and what is being done to prevent them? These are the compelling questions that we address in this chapter.

First, a disclaimer of sorts. It is unrealistic to consider intercity passenger rail security issues apart from the consideration of the general rail system's security. As noted in the chapters on rail freight and commuter rail security, the system is interconnected in so many ways that policies and procedures for security passenger rail must be seen in light of its connections – physically, organizationally, technically – to the overall North American rail network. Nonetheless, there are specific threats, vulnerabilities, and responses that make intercity passenger rail security a topic worthy of consideration on its own. In the U.S., as a prime example, security measures adopted for passenger airlines have served as only a partial model for passenger rail to follow. To many travelers accustomed to the rigid security measures adopted for airline travel, intercity rail travel is a pleasant alternative, harking back to the more untroubled travel of the pre-9/11/01 era.

Second, as is so often the case regarding rail security, it is difficult to separate measures designed for safety considerations from those designed specifically to prevent and react to terrorist activities. We do not dwell on some of the specific measures advanced recently to enhance rail safety, such as PTC or rules preventing train crews from texting or making personal cell phone calls while operating trains. However, there are significant ways in which calls for greater safety measures draw attention to and complement current security measures.

We begin our review with a brief history of attacks on passenger rail around the world. We then describe the current state of intercity passenger rail in North America, including in particular the rise of the "intercity corridor" as a major focus for investing in new or improved passenger rail service. The Northeast Corridor is treated as a separate example, given its high density of passenger operations, its linking of cities that are at the top of the list of proven and likely terrorist targets, and with ownership of the most trackage in the hands of Amtrak.[1]

With the descriptions in place, we move to relate historical examples and current operations to identify the nature of threats and vulnerabilities to intercity passenger rail; explore the policies and programs established to secure passenger rail and the responsibilities assigned to various organizations that are involved in the various aspects of passenger rail security; and suggest what appear to be the most realistic options to enhance security efforts in the future.

Terrorism and passenger trains: an international review

The review of terrorist attacks worldwide on passenger rail reveals several disturbing but highly relevant insights. First, attacks on rail systems – trains and stations – has been on the rise since the late twentieth century, fueled by the rise of radicalism both within nation-states and internationally. Second, security measures have been inadequate to prevent attacks both onboard trains and in supporting rail infrastructure. Third, most attacks have been attempted using traditional means, explosive devices or mass shootings. Fourth, the goal of terrorist attacks has been to kill passengers randomly and create fear in the mind of the public when using passenger rail. Terrorist events have created secondary costs as well, such as disrupting freight as well as passenger service or causing economic chaos in cities experiencing a major event, but it is not clear if anything other than the desire to kill and create fear among the populace, in revenge for real or supposed grievances, has factored in to the thinking of those executing catastrophic rail attacks.

Significant terrorist acts against passenger rail have occurred in Europe, the Middle East, South Asia, and Africa. Most have been attacks on commuter trains or urban transit, including the Madrid, London, and Mumbai attacks since 2004 that have garnered the most attention and have been important examples used by U.S. rail operators and policy makers to draw lessons and develop risk-based policies. However, a number of examples of attacks on intercity passenger trains have occurred, including:

- In Italy on December 23, 1984, 16 people died and over 200 were injured when a bomb exploded on a Florence to Rome train. This followed the bombing earlier in 1984 of the Bologna Central Station that killed 85. The bombings were attributed to a neo-Nazi terrorist group.
- Between the towns of Zenza and Dondo, Angola, rebel forces derailed a train and killed 252 passengers by gunfire.
- In separate attacks in 2002, Maoist guerrillas in North Central India killed over 140 passengers.
- The Nevsky Express, the premier passenger train between Moscow and Saint Petersburg, was derailed by an IED planted under the rails, causing 27 confirmed deaths and roughly 100 injuries. Terrorism was blamed for the crash, with the exact terror group unclear but likely based in the North Caucasus region of Russia.
- A passenger train was derailed in West Bengal, India, on May 28, 2010, and hit by an oncoming freight train, killing at least 65 passengers.

Attacks on intercity trains are far less common than attacks on commuter or urban transit trains, or on rail stations. Commuter lines and transit offer greater opportunities for mass casualties and also disrupt the normal activities of a city. Both they and stations, despite enhanced security efforts, remain largely "soft targets" for terrorists, as discussed in the chapter on commuter rail.

The review of terrorist attacks on intercity trains also indicates that the source of terrorism is in every instance domestic terrorism, sectarian or ethnic strife, or regional animosities (e.g. India and Pakistan, or Azerbaijan and Armenia). The spread of global jihadism, nurtured by terrorist organizations with global ambitions and aided by advances in communications such as social media, may nonetheless be seen as the most likely emergent threat to intercity rail, as it is for other targets of terrorism.

Amtrak and the national intercity rail system

Since its inception on May 1, 1971, Amtrak – the common name for the National Railroad Passenger Corporation – has operated a national system of passenger trains, including long-distance intercity passenger trains, intercity corridor trains, and, by contract with state and local entities, commuter trains. At first, four private railroads – Southern Railway; Denver, Rio Grande & Western; Chicago, Rock Island & Pacific; and Georgia Railroad – continued to run intercity passenger trains, but as of 2016 only the state-owned Alaska Railroad operates intercity passenger trains besides Amtrak.

Amtrak's history has been well documented elsewhere. Suffice it to say that many observers of the railroad scene, even well informed and sympathetic ones, would be surprised to see that Amtrak has survived and in many respects prospered since its formation in the early 1970s. As Amtrak notes in its FY2015 National Fact Sheet, ridership continues to climb, with over 30.8 million total passenger trips logged between October 2014 and September 2015, the fifth consecutive year with ridership over 30 million. An average day sees more than 300 scheduled trains with 84,600 passengers served.[2]

Amtrak's success can also be seen in the resurrection of major passenger stations such as Washington Union Station in the District of Columbia and Philadelphia's 30th Street Station; in its development of partnerships with states to bolster its resources and add route miles and enhance service on intrastate and interstate urban corridors; and, most crucially, in its attempts to fend off efforts to defund it, privatize its most critical routes, or in other ways fundamentally change its role as the single national provider of intercity passenger rail service.

At its start, Amtrak operated by contract on rail mileage owned by so-called "host roads": privately owned freight railroads that were happy to pass off money-losing passenger operations to the new corporation. This changed on April 1, 1976, with the acquisition of the Northeast Corridor (discussed in more detail below) as part of the formation of Conrail. Today, Amtrak continues to operate over 70 percent of its route miles under contract with host railroads (including some publicly operated regional commuter lines) and the major freight railroads.

Since the acquisition of the Northeast Corridor (with only a few segments state owned), Amtrak has also acquired the 104.2-mile Keystone Corridor (between Harrisburg and Philadelphia) and the Michigan Line (a 95.6-mile route between Porter, IN, and Kalamazoo, MI).

The map in Figure 9.1 illustrates the intercity passenger routes operated by Amtrak. Amtrak has identified 34 different "services" on its website. Several of the services operate over the same route miles as other services; examples include the Acela and Northeast Regional services, which both utilize the busy tracks of the Northeast Corridor, as do trains operating from New York to various locations in the Southeast, Middle Atlantic, and Midwestern states.

The map also shows that north/south routes dominate along both coasts, with east/west routes more common in the Midwest, such as the Sunset Limited from New Orleans to Los Angeles. Long-distance trains fan out from Chicago to Texas, California, and the Pacific Northwest. Between central Texas and California, there are no north/south routes, reflecting the traditional east/west travel in the Rocky Mountain and Great Plains regions.

Routes and train frequencies are roughly explained by two factors: population density and local and state support for passenger rail. Virtually all the major population centers of the U.S. have some intercity passenger service; although outside the major corridors (to be discussed below) train frequency is typically low. Over 500 communities are served by Amtrak-operated trains, from major cities like New York, Los Angeles, and Chicago to small station stops like Parkesburg, PA, Pauls Valley, OK, and Thurmond, WV.

Roughly half of the services included by Amtrak have been created through funding by states, usually through state departments of transportation or public authorities created by state action. States have partnered with Amtrak to provide service on routes abandoned with the establishment of the first Amtrak system map in 1971, or to upgrade services by increasing train frequency or infrastructure on regional corridors already serviced by Amtrak, such as Pennsylvania's Keystone Corridor. In several cases, adjoining states have partnered to provide interstate service, such as the Milwaukee-Chicago Hiawatha corridor between Wisconsin and Illinois, and the Maine, New Hampshire, and Massachusetts backed *Downeaster* service between Boston and Brunswick, ME.

Passenger corridors, whether intrastate or interstate, have several common features. They are typically middle range in terms of distance, usually from 200 to 500 miles. They typically offer more than once-per-day service, may not require seat reservations, and have large anchor cities at each end (or extend beyond an anchor city to the far end of a metropolitan area, as in the case of California's Capitol Corridor). Population density within the corridor and the distance between logical anchor metro areas, combined with the political viability of corridor investment, helps explain why most of the corridors are in the heavily populated Northeast, Pacific Coast, and Great Lakes regions.

While some corridor passengers use the trains as long-distance commuter trains, the attraction of the corridors resides in their comparative market advantages vis-à-vis their modal competitors, passenger airlines, and automobiles, for a

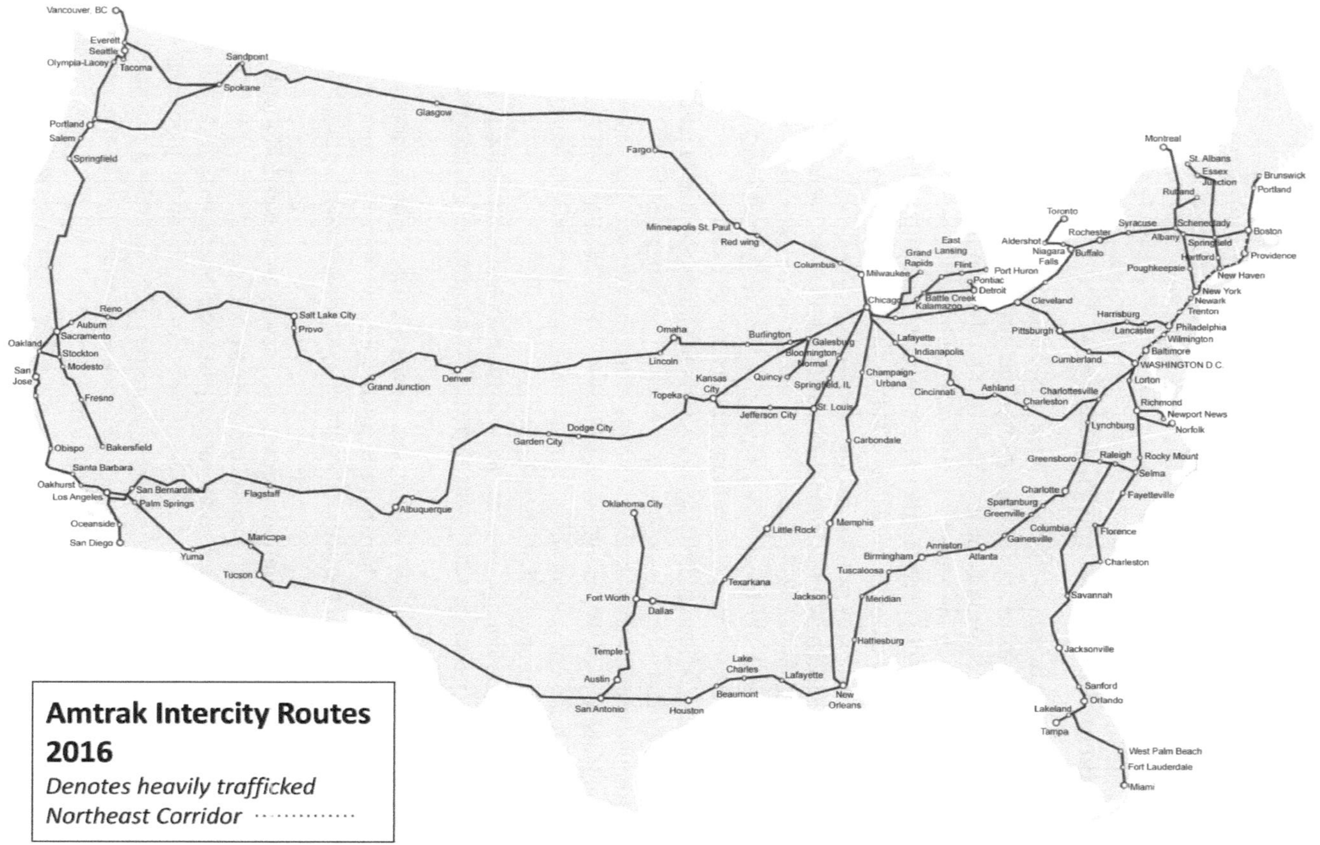

Figure 9.1 U.S. intercity passenger rail routes

Source: adapted from Wikipedia and Alaska Railroad

variety of travel reasons. Easy entry and exit means less wasted time at airports, and destinations in downtown locations means avoiding traffic congestion and delays in urban settings. For states strapped for resources to build more roads, or where urban development means a lack of appropriate land for expanding the road system, or where public sentiment is pro-rail and anti-highway, the partnership with Amtrak, while not always trouble-free, has seen a remarkable growth in the availability of rail passenger options compared to the railroad map of 1971 at the start of Amtrak.

The Northeast Corridor: a special example

The Washington to Boston Northeast Corridor is the jewel in the Amtrak system, and the example that makes the strongest argument for investing in intercity passenger rail in the U.S. Amtrak owns 363 of the total 457 miles between Washington and Boston, with the remaining miles owned by public sector entities: the New York Metropolitan Transportation Authority, the Connecticut Department of Transportation, and the Commonwealth of Massachusetts. The route, however, is run as a seamless whole by Amtrak. Service is provided by both high-speed Acela trains and by electric-powered regional trains. In addition to intercity trains, a large volume of commuter trains share the same tracks and terminals, as do some freight trains. Over 2,200 trains of all types use the Northeast Corridor tracks daily, making it the most densely trafficked rail line in the nation.

The Northeast Corridor intercity service shows the inherent value of passenger rail in a corridor that is heavily populated; has large population centers not only at its outer boundaries but also at several points along the corridor (e.g. Baltimore, Philadelphia, Newark (NJ), as well as Washington, New York City, and Boston); and provides more convenient downtown-to-downtown service than competing modes, especially commercial airlines.

The Northeast Corridor also provides the greatest risks and vulnerabilities of any part of the national passenger rail system. Serious risk factors include at least these:

- Washington, as the nation's capital, and New York City, as its major population center and financial hub, have been shown to be primary targets of terrorism in the past, and continue to be seen as symbols of American power and national identity.
- The sheer volume of train movements and passengers makes it difficult to monitor and check for suspicious activity at stations.
- The mingling of commuter and intercity passenger operations means entry and exit in the corridor occurs at a large number of stations and to trains operated by commuter rail systems as well as Amtrak. This has created a complex network of organizations involved in policing and security, and a large number of access points that require attention.
- The corridor includes several locations where terrorist events can wreak special havoc, including crowded station areas with underground trackage, and tunnels at Baltimore and under the Hudson River approach to New York City.

In addition are large numbers of bridges and other infrastructure that can create significant targets for terrorist activity.

- The higher speeds of intercity trains makes targeting of trains at speeds of 100 miles per hour or more likely to cause more catastrophic results than slower-moving trains.
- The corridor moves past the New Jersey Chemical Coast and the Delaware Chemical Cluster, two of the major sources of hazmat in the U.S.

In short, if we assume that terrorist attacks are more likely on passenger trains than freight trains, and that they are likely to be staged to create the most casualties and disruption to service, and to be effected in the most symbolic locations, the Northeast Corridor stands out as the greatest challenge to all those agencies working collaboratively to prevent such events from occurring or, if unpreventable, to develop ways to most effectively respond and mitigate the damage wrought.

Public policy to address threats and vulnerabilities

Looking at the development of public policy for passenger rail security, three characteristics jump out: it has been *incremental*, building gradually through legislation, executive action, and industry action; *reactive*, with occasional bursts of attention after international events such as the Madrid train bombings in 2004 and the London Metro attack in 2005 (and to a certain degree, events on other transportation modes that might be replicated on passenger trains or stations); and *collaborative*, involving federal agencies such as the TSA and FBI, state and local police and homeland security agencies, and rail operators and the associations that represent them and help provide a common basis for action. For intercity passenger rail, the key players have been Amtrak, DHS, to include TSA, the FRA, and police departments in the major cities served by passenger rail, as will be noted below.

Policy development has also been heavily influenced by research into the past examples of attacks on passenger rail and by modeling or envisioning likely threats and vulnerabilities and their probable consequences. Given the dearth of empirical examples from the American experience on which to build, these efforts have been guided either by analysis of international examples or by risk analysis, based on threats and vulnerabilities of the passenger rail system and likely consequences of possible events. At the forefront of such work has been both public agencies such as the GAO, TSA, and committees of Congress and the Congressional Research Service (CRS), and private or university-based think tanks such as RAND Institute and the Mineta Institute at San Jose State University in California.

While passenger rail was considered a possible target in the days following the 9/11/01 attacks, it was several years before a systematic assessment of threats to passenger rail began to emerge. Although some studies predated the Madrid attacks of March 2004, that event and the 2005 London Metro bombings led to an increased fear that similar attacks might strike the U.S.

In a 2005 report to Congress entitled *Passenger Rail Security: Federal Leadership Needed to Prioritize and Guide Security Efforts* (GAO 2005a), the GAO recognized the reactive and incremental nature of actions to secure the passenger rail system. Noting that the Madrid and London bombings had "dramatically highlighted the vulnerability of passenger rail systems worldwide to terrorist attacks and the need for an increased focus on security for these systems" (p. 5), the GAO research team looked at both domestic and foreign passenger rail operators to review actions taken to assess risks and enhance security, and they compared actions taken in different national contexts. The resulting study was at the time the most comprehensive review of security risks involving passenger rail systems. As is common, the study included commuter rail and intercity rail, as well as urban transit systems, within the report.

The GAO reviewed the actions taken since 9/11/01 and particularly after the Madrid attacks in May 2004. It found that TSA moved quickly and without the normal public comment period in issuing security directives for rail operators to conduct frequent station inspections and to utilize K-9 detection teams. Operators criticized the TSA for the vagueness of these directives. In addition, discrepancies were identified between existing safety rules promulgated by the FRA and the new TSA rules, showing again the need for close collaboration between the various entities charged with rail security. This resulted in a Memorandum of Understanding (MOU) between the two departments as suggested by GAO.

GAO generally found similar approaches to passenger rail security between the U.S. and foreign rail operators and similar risk factors. These were summarized as follows:

> Specifically, most U.S. and foreign operators we contacted had implemented customer awareness programs to encourage passengers to remain vigilant and report suspicious activities, increased the number and visibility of their security personnel, increased the usage of canine teams to detect drugs and explosives, enhanced employee training programs, upgraded security technology, tightened access controls, and made system design improvements to enhance security.
>
> (GAO 2005b, p. 8)

GAO tries to avoid blaming federal executive branch agencies for faults, and it favors a constructive approach to improve policies and programs. However, it is clear in reading the report that TSA needed to work closely with stakeholders to enhance cooperation, speed up actions, and incorporate risk-based analysis. It also found that foreign examples and practices are relevant to such efforts. The report concluded by providing recommendations for executive action. This included the need to speed up and refine the use of risk analysis by TSA and the Office for Domestic Preparedness within DHS, and charge TSA in collaboration with DOT and the passenger rail industry to develop industry-wide standards to be used by TSA inspectors, set timelines for completion of MOUs between participating organizations, and enhance ways of sharing information.

A report titled *Detour Ahead: Critical Vulnerabilities in America's Rail and Transit Security Programs*, prepared by the Democratic staff of the Committee on Homeland Security of the U.S. House of Representatives, made many of the same recommendations as the GAO, albeit with a more critical tone regarding the efforts to date by the Bush administration. The report highlighted the need for increased funding for rail security and the need for TSA to develop an overarching plan to guide rail security efforts. Intercity passenger rail was not singled out for specific recommendations in the study.

Surface transportation security, however, still accounts for a tiny portion of the resources and attention given to transportation security in the U.S. The most telling statistic in this regard is that 97 percent of TSA's resources goes to airline security. Of course, this figure reflects the large personnel budget for air due to TSA's employment of airport security screeners. But it also illustrates how salient past air attacks, most notably the attacks of 9/11/2001, remain in guiding national security efforts.

TSA and the rail passenger

Some of TSA's surface programs – in particular, railroad programs – have been discussed previously. But what precipitated this heightened awareness and involvement of the federal government in protecting passenger railroad operations you may ask? The 2004 Madrid train bombings were the watershed event.

Subsequent to the Madrid train bombings, the president and Congress realized that what happened there could happen in the homeland. Therefore, the Transportation Security Administration was directed to create the Surface Transportation Security Inspection Program (STSIP) in 2005 to address operational security for passenger rail and rail transit systems nationwide. This was accomplished by hiring 100 transportation security inspectors, furthering partnered initiatives with the industry and issuing two security directives (SD)[3] that required protective measures be taken: one addressing passenger rail operations conducted by passenger rail operators, to include commuter rail, and the other Amtrak and Alaska Railroad passenger operations. The "first 100" inspectors and, in particular, its regional leadership was staffed with individuals with significant railroad and transit experience to ensure industry credibility.[4] The protective measures of the SDs focused on designating security coordinators, reporting threats and significant security concerns, notifying personnel of security concerns, station and facility inspections, higher visibility at stations, to include K-9 teams and vulnerability assessments – all of which became the basis for partnered security measures and regulation. Both SDs included the freight railroads over which passenger trains operated.

As the STSIP evolved, it grew to over 300 inspectors in nearly 50 offices in most of the lower 48 states and Alaska and Hawaii. The expansion was based on, among other things, the enactment of 49 CFR 1580, Rail Transportation Security, and the operation T/PIH trains through UASI[5] designated high threat urban areas (HTUA), new and expanded commuter rail operations and transit systems, and where history has shown security to be a concern. As the program evolved, the focus of the STSIP expanded to include freight rail other than just sharing track with passenger operational, and in particular the handling of hazmat; intermodal and highway motor carrier; intercity and transit bus; pipelines; and the landside of maritime. This growth also included partnering with the FRA and other federal agencies and industry groups, such as the Association of American Railroads (AAR) and American Public Transportation Association (APTA).

TSA includes intercity passenger rail as one element of mass transit for purposes of programming and mission development. In practical terms, this means that Amtrak, as the operator of intercity rail nationwide, is eligible for funding for security grants and is included in programs designed to assess risks, provide equipment and training, and share information to enhance security. As is the case for all transportation modes other than air, TSA, as explained by former TSA John S. Pistole, "does not conduct frontline screening, TSA engages with state, local, and private sector partners to reduce vulnerabilities, assess risk, and improve security through collaborative efforts" (Pistole 2014, p. 5).

TSA's role in providing security for all modes of transportation, not just air, has been a part of its mission since its formation and was reinforced by specific charges from Congress in the Implementing the 9/11 Commission Report Act of 2007. Illustrating both the incremental and the collaborative nature of surface transportation security policy that pertains to passenger rail are several relatively new programs supported by TSA: the BASE program; Operation RAILSAFE; and the Intermodal Security Training and Exercise Program (I-STEP). In addition to these initiatives, Amtrak has participated for a decade in TSA's Transit Security Grant Program (TSGP), which provides funds for a variety of ways to enhance security. Each will be discussed below.

The Baseline Assessment for Security Enhancement (BASE) program is part of TSA's effort to develop risk analysis and assessment as the basis for all its programs and operations. It focuses on 17 security action items developed in a collaborative effort between TSA and the Federal Transit Administration and is a "snapshot in time" of a rail passenger entity's security programs and processes. In the BASE program, TSA Transportation Security Inspectors – Surface (TSI-S) assess corporate security efforts at mass transit, Amtrak, and large intercity bus operators.

Operation RAILSAFE is one of the newer programs and builds on the same emphasis on metrics and analysis as the BASE, with a specific application to passenger rail (Pistole 2014).

TSA and Amtrak have a long-standing security partnership through programs that aim to deter terrorist activity through expanded random, unpredictable security activities. The first of two is the TSA's Visible Intermodal Prevention and Response (VIPR) program. The VIPR program is intended to enhance the security at transportation facilities to create a randomness and scale of security presence and force structure so as to deter or detect the planning or execution of a terrorist attack. It is a program of TSA's Office of Law Enforcement (Federal Air Marshals) in conjunction with the Office of Security Operations, Office of Security Policy and

Industry Engagement, and federal, state and local law enforcement agencies. With Amtrak, the VIPR focus is on railroad stations but includes infrastructure and facilities of the other modes of transportation, to include aviation.

The second program and as stated previously is the Regional Alliance including Local, State, and Federal Efforts (RAILSAFE). It is an expansion and evolution of the VIPR concept and coordination with rail and transit agencies and local law enforcement and is intended to enhance the coordination among the agencies to further counter-terrorism activities, such as heightened station and right-of-way patrols, increased security presence on board trains, explosive detection K-9 teams sweeps, and random passenger bag inspections. On average more than 40 states and over 200 agencies participate in TSA's VIPR teams and industry-sponsored RAILSAFE programs (Pistole 2014, p. 5).

As TSA's surface program grew, it was quickly learned that the railroads and transit systems, which already were experienced in security of their operations, infrastructure, and facilities, preferred partnered initiatives rather than regulations. This in part was due to the fact that regulations often came with unfunded mandates, and working with the government, in a partnered and cooperative environment, was more productive and less costly. This approach was enhanced because TSA had the foresight to hire railroad and transit folks in the "first 100". As the STSIP and OSPIE evolved and worked with the industry, proactive security measures complemented the operational and reactive security postures typically fostered by TSA. Partnered with other DHS organizations, such as CBP, and federal agencies, such as the FRA and FTA, a coordinated, communicated and comprehensive partnership materialized with the industry and is being furthered to enhance a truly multi/intermodal approach to transportation security.

Conclusions: where do we stand, and where do we go?

Intercity passenger rail security rests on a paradox: it has not been the target of terrorist activities since the 1995 Sunset Limited derailment, but it continues to be a remarkably soft target for potential attacks. Traditional means of prevention and detection have been enhanced since the 2000s, but security on the nation's intercity trains lags far behind that of intercity air travel. To many, this is the advantage of using the train: no long security lines, no taking off belts and shoes, no restriction on beverages, etc. A single attack leading to mass fatalities, however, might require a tightening of security along the lines of airport security. This flies in the face of the openness and multiple access points of the Amtrak system: what might be feasible (if onerous) security arrangements at Penn Station, Manhattan or Union Station, Washington, would be costly and difficult to engineer at the numerous smaller entry points of the system. A former Amtrak Deputy Police Chief said that "their concern was for a person or persons with backpack IEDs getting on a train at an outlying station with minimal security, setting the timer and getting off a station or two before a main terminal." An example would be a person getting on a MBTA commuter rail train at Bridgewater on the Middleboro/Plainville line and getting off at either Quincy Center or JFK/UMass just before

South Station terminal, which is shared with Amtrak, the MBTA Red and Silver Lines, and privately-owned intercity bus lines.

The joint usage of tracks and stations for commuter as well as long-distance trains makes it impractical and likely impossible to screen passengers through airport-type restricted entry, given that, during peak hours of operation, commuter rail combined with transit and intercity passenger rail generates more passengers than aviation. Moreover, the unsecure area of the station or terminal is far larger than the platform area and cannot be secured using an approach similar to those of airport terminals. Greater use of other methods of enhancing security, such as CCTV and other monitoring technology, used in combination with human monitoring through programs such as VIPR and RAILSAFE, are ways to enhance surveillance without hindering entry and exit. Physical design using the principles of CPTED[6] can assist in maintaining security by eliminating hiding places, creating open lines of sight, and using proper lighting. Security is enhanced by a combination of measures and not just one.

Notes

1 Amtrak has about 21,000 route miles with less than 10 percent owned by them. The Northeast Corridor is a well-known major passenger route but is not Amtrak's only one.
2 By comparison the Bureau of Transportation Statistics reports that 2016 saw slightly more than 20 million daily domestic airline enplanements.
3 A security directive is an interim measure addressing security threats, issues, or concerns that span the gap until the threat, issue, or concern is resolved or a regulation is put in place. An example is the initial handling of the 300-foot parking ban at airport terminals in response to the Glasgow, Scotland, terminal bombing in 2007.
4 Of the "first 100", there were 12 regional directors with over 300 years of operating railroad and transit system experience collectively.
5 The Department of Homeland Security's Urban Area Security Initiative (UASI) is a program that provides funding to enhance regional preparedness and capabilities to enhance terrorism prevention activities in designated high-threat, high-density areas.
6 Crime Prevention through Environmental Design (CPTED) is a multi-disciplinary method of deterring criminal behavior, vandalism, and terrorism by using strategies and measures that deter the planning or conducting of a criminal or terrorist activity.

Bibliography

Kaplan, E. (2007). *Rail security and the terrorist threat.* Council on Foreign Relations. Retrieved from www.cfr.org/publication/12800/rail_security_and_the_terrorist_threat

Peterman, D. R. (2005, January 24). *Passenger rail security: Overview of issues* (Congressional Research Service Report for Congress No. RL32625). Washington, DC.

Pistole, J. S. (2014). *Development of TSA's risk-based approach to security.* Retrieved from www.tsa.gov/news/testimony/2014/04/30/development-tsas-risk-based-approach-security

Rai, S., and Sengupta, S. (2006, July 12). Series of bombs explode on 7 trains in India, killing scores. *New York Times*. Retrieved from www.nytimes.com/2006/07/12/world/asia/12india.html

Rail Safety and Standards Board. (2009). *Report on improvements in the safety of passengers and staff involved in train accidents.* Retrieved from www.rsb.co.uk

Rosenbloom, S. (2003). New departures: Rethinking rail passenger policy in the twenty-first century. *Journal of the American Planning Association*, 69(2), 203–204.

Schenk, F. (2010). Attacking the empire's Achilles Heels: Railroads and terrorism in tsarist Russia. *Jahrbucher Fur Gesschichte Osteuropas*, 58(2), 232–253.

Sneider, J. (2015). With rail security planning, information sharing is key. *Progressive Railroading*. Retrieved from www.progressiverailroading.com/security/article/With-rail-security-planning-information-sharing-is-key-43156

U.S. Congress (2015). *H.R.749-Passenger Rail Reform and Investment Act of 2015*. Retrieved from www.congress.gov/bill/114th-congress/house-bill/749

U.S. Department of Transportation (2009, October). *Recovering from disasters: The national transportation recovery strategy*. Retrieved April 29, 2017 via https://www.transportation.gov/policy-initiatives/disaster-recovery/recovering-disasters-national-transportation-recovery-strategy

U.S. General Accountability Office. (2004, March 23). *Rail security: Some actions taken to enhance passenger and Freight rail security, but significant challengers remain*. GAO-04–598T.

U.S. General Accountability Office. (2005, October 4). *Amtrak management: systemic problems require actions to improve efficiency, effectiveness, and accountability*. GAO-06–145.

U.S. Government Accountability Office. (2005a, September). *Passenger rail security: Enhanced federal leadership needed to prioritize and guide security efforts*. GAO-05–851. Retrieved February 22, 2006 from Government Accountability Office via www.gao.gov/htext/d05851.html

U.S. Government Accountability Office. (2005b, October 20). *Passenger rail security: Enhanced federal leadership needed to prioritize and guide security efforts*. GAO-06–181T. Retrieved February 22, 2006 from Government Accountability Office via www.gao.gov/htext/d06181t.html

U.S. Government Accountability Office. (2006, March 29). *Passenger rail security: Evaluating foreign security practices and risk can help guide security efforts*. GAO-06–557T. Retrieved December 7, 2006 from Government Accountability Office via www.gao.gov/htext/d06557t.html

U.S. Transportation Security Administration. (2015). *TSA supports passenger rail systems security*. Retrieved from www.tsa.gov/news/top- stories/2015/09/03/tsa-supports-passenger-rail-systems-security.

Waugh, W. L. Jr. (2008, Winter 2008–09). Assessing the risk to rail and transit systems. *The Public Manager*, 71–76.

10 Emergency or incident response, operations, and security

The division engineer of a major railroad on the Gulf Coast is preparing for a Category 3 hurricane aimed to make landfall between Houston and New Orleans. The National Oceanographic and Atmospheric Administration (NOAA), National Weather Service (NWS), has forecasted landfall in 120 hours as shown in Figure 10.1, a typical five-day storm track.

The engineer immediately began to think of the 150 toxic/poison inhalation hazard (T/PIH) rail cars in the yard and how the railroad must either evacuate them to a safe area or protect them in place. Others at the railroad also began to reflect on a recent DHS message stating that terrorists may take advantage of landfall, when everyone else would be hunkering down, to attack or sabotage the T/PIH rail cars. Still others wondered about the Amtrak arrivals and departures in the period before, during and after landfall?

Management called a meeting of key resource protection, operating, and maintenance personnel to assess the potential impacts and determine what courses of action to take. They first determined where to evacuate the T/PIH and other hazmat rail cars and, if all cannot be evacuated, how to protect the ones remaining. They then looked at protecting the signal, communication, and power systems and insuring that the railroad would have backup power, should it be needed. Other team members contacted Amtrak and advised that all passenger train service in the area would be halted by 24 hours from landfall. They negotiate with them to evacuate some nonessential personnel and to work with the local government to use Amtrak to evacuate those people who do not have the capability to do so. Once all is accomplished, senior management itself relocates to a safe location so that once landfall is over, it can immediately begin to assess the condition of the railroad and its environment.

As everyone is hunkering down, many had thoughts of who they would coordinate with after landfall. The list was a long one: TSA and the Federal Railroad Administration with regard to railroad safety and security as operations resumed; local law enforcement to augment the railroad police to ensure that no criminal or terrorist activity has occurred during landfall; and how to expeditiously implement the inspection of the physical facilities as detailed in the regional continuity of operations plan (COOP) for the company. Finally the team members further contemplated what additional resources might be needed and how to back up those staff that may have been impacted by the hurricane.

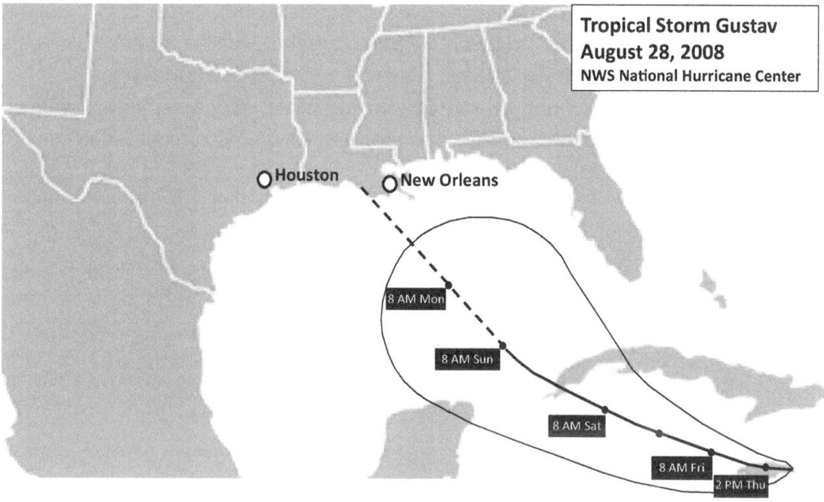

Figure 10.1 Typical NOAA/NWS five-day track

Source: National Oceanographic and Atmospheric Administration

This scenario can be played out for numerous natural and human-caused disasters: tornados in Oklahoma, earthquakes in California, wildfires in Colorado, and blizzards in Colorado and New England. An *all hazards plan* will employ standard procedures for all disasters but contained appendices added for the specific disasters in a given area. For manmade disasters, which is to include terrorism, the *all hazards plan* addresses the common processes and unique approaches for, among other possibilities, a targeted attack on a hazmat rail car or train, simple copper theft in a rail yard, or placing an IED on a railroad bridge over a busy highway. Other manmade disasters that could occur are a hazmat spill along the right of way or failure of a coupler causing a derailment, which could have a wide range of impacts and involve an equally wide range of response assets.

Overview

The discussion in this chapter will focus on Gulf Coast hurricanes, but the information, processes, and procedures can be extrapolated for any incident involving railroads in the U.S. The *all hazards approach* provides the foundation for disaster planning, preparedness, response, recovery, and mitigation. Where the planning will differ is for the regional and geographic specific natural and manmade disasters.

When Hurricane Rita approached the Gulf Coast, the natural tendency of the freight railroads was to evacuate the hazmat and in particular the T/PIH rail cars. Hence, based on the hurricane track, the railroads began the evacuation west and inland towards Houston and beyond. To the railroads' dismay, the yards and branch lines in and around Houston, and to some degree the mainlines, were at or near capacity because of Hurricane Katrina a month earlier. Thus, many hazmat and T/PIH rail cars had to be protected in place to whatever extent possible. This occurred even though there was advance warning via hurricane tracking.

Evacuating and/or protecting all rail cars is important, and, even with advanced warning, it was difficult because of the proximity in time of Hurricanes Katrina and Rita: within 30 days of each other. Take the problems that occurred and resulting risk, and couple it with an event with little or no warning, such as a tornado, earthquake, or flash flood – the problem is exacerbated. The challenge becomes great. Other problems exist with blizzards, snow melt and associated flooding, and wildfires.

The railroads and the problems experienced with Super-Storm Sandy were similar to that with Hurricanes Katrina and Rita. But the problems were exacerbated by the extensive commuter and passenger rail systems in the Northeast.

Introduction

For the purpose of this chapter, the focus will be on freight and passenger rail, to include both the commuter agencies and Amtrak. Although rail transit has similar issues and concerns, for the most part they do not operate in the same corridor and/or on the same track. Therefore, the risks associated with the proximity of freight and passenger rail operations are not a major concern.

When considering security during emergency or incident response, one must first determine what constitutes an emergency or an incident. The FTA's Security and Emergency Planning Guide essentially defines an emergency as "a situation that is life threatening or causes significant damage or disruptions that reduce the ability to operate" (Federal Transit Admin 2003). FEMA defines an incident as "an occurrence or event, natural or human caused, that requires an emergency response to protect life and property." FEMA further defines an emergency as "any incident that requires a response to protect life and property" (FEMA 2008).

When looking at security in emergency planning, preparedness, and response, one must first look at the types of disasters, emergencies, and incidents that are to be considered and, in the case of terrorism, the intent to cause harm equal to or beyond that caused by a natural disaster. Disasters are natural (such as hurricanes and tornados) or manmade (such as an accident, vandalism, or terrorism). The difference among accidents, vandalism, and terrorism is intent. An accident is unintended, but vandalism and terrorism have intent. Vandalism, as defined in the *Merriam-Webster Dictionary*, is the willful or malicious destruction or defacement of public or private property. The intent of vandalism is generally mischievous. The *Merriam-Webster Dictionary* defines terrorism as "the use of violent acts to frighten the people in an area as a way of trying to achieve a political goal." The difference between vandalism and terrorism is the ideology involved, magnitude of the event, and intended outcome, to include fear.

In emergency planning and in the reconstitution of operations, security awareness is important. This is because individuals or organizations could take advantage of the railroad when hunkering down, during landfall, and immediately after to disrupt or compromise railroad operations and infrastructure. It could also include compromising weaknesses in operations and critical infrastructure, such as bridges and tunnels.

A couple of cases in point were during Hurricanes Rita and Ike. Prior to landfall during Hurricane Rita, a security advisory was circulated because of "chatter" involving potential terrorist activity in and around rail yards when all railroad personnel had found appropriate shelter. This required the railroads along the Gulf Coast to take extra measures to secure its infrastructure and equipment. After Hurricane Ike, one railroad had many of its portable generators that powered grade crossing signals stolen. Although probably stolen for need, the potential existed that they could have been stolen to cause an accident between a train and a vehicle, but it also stretched the resources of the railroad to monitor the locations and replace the stolen generators. Could this provide an opportunity to compromise a bridge, tunnel or grade crossing?

When looking at emergency planning and response, many federal and state agencies, transportation providers, and industries are involved. For Gulf Coast hurricanes, and along the Texas Gulf Coast area in particular, there is a significant population of involved organizations. At the federal level, FEMA is the lead agency, but they are supported by TSA and the Coast Guard. The FRA is also involved, as are the other modal agencies of the U.S. Department of Transportation. At the state level, the department of emergency management, state police, and department of transportation are some of the involved agencies, as are the regional and local equivalents of the federal and state agencies. Chemical plants along the Gulf Coast, the I-95 (New Jersey Turnpike) corridor, and other locations that involve the transportation of hazmat to/from their facilities are active participants. For major disasters or incidents, the FEMA Incident Command System (ICS) is set up. All involved parties, to include industry and volunteer groups, play a role in or are involved with ICS. For example, during Hurricane Katrina, a FEMA Joint Field Office (JFO) was set up in Baton Rouge and remained operational *several years* afterward.

FEMA's National Response Framework (NRF) has four mission areas. They are prevention, protection, mitigation, and recovery. The NRF continues to be the overarching principles that frame the "boots-on-the-ground" approach to determining the phases of an incident or a disaster. The phases of the mission are commonly referred to as the *emergency management cycle* and specifically include preparedness, prevention, mitigation, response, and recovery.

By proper planning and preparation, railroads and transit systems can minimize the risk of an event and maximize the protection of their assets by learning from past experiences moving forward. *Prevention* is the act of proactively protecting a railroad's assets, to include equipment, infrastructure, and its people. It answers the question of what can a railroad do to reduce vulnerability, deny access to its facilities, and deter wrong doing. *Mitigating a threat* is taking the thought process developed during planning and preparedness and reducing the threat and accompanying risk by physical and/or operational means. Proper *response* to an event takes prior planning and coordination among all parties, to include law enforcement, first responders, and government agencies at all levels, so that access to the affected area is possible with minimal delay and people and property are protected. Communications interoperability is important for this to happen. This was brought to light in emergency management with the inability of emergency operations groups, first responders, law enforcement, etc. to effectively communicate during Hurricane Katrina. Finally, delays in the resumption of service could further compromise the *recovery* effort and increase the exposure potential by the further idling of those hazmat rail cars that had earlier been evacuated or protected in place. Properly staged and positioned resources, material and equipment are essential to expedite the recovery, protect assets, and resuming normalcy.

The remainder of this chapter will focus on natural disasters and Gulf Coast hurricanes, in particular, but could be applied in principle to terrorist attacks and vandalism. Hence, an objective discussion can be achieved that links security with emergency preparedness, rather than a theoretical one.

Phases of an emergency or incident

As you plan for future disasters impacting your railroad, you wonder what resources you should consider and how to phase your plan. You reference the FEMA National Incident Management System (NIMS) to assist you to understand how to manage incidents given the threats and vulnerabilities that could affect your railroad and its operations. You review NIMS and recognize that your planning should cover the phases of a disaster. In accordance with NIMS, you recognize that your disaster or emergency operations plan must reflect an *all hazards approach*[1] to a disaster and include planning for it, how you prepare for it, what to do during the disaster (i.e. landfall for a hurricane), and how you will respond and recover from it. Your work is cut out, as there are several natural and manmade disasters, to include terrorism, which you must consider. Along the Gulf Coast, hurricanes, oil spills, petrochemical plant incidents, and flooding are some of the disasters you should consider. You think, what if a hurricane impacts an oil rig in the Gulf of Mexico or damages a petrochemical plant exacerbating things? This combination of a natural and technological disaster is commonly referred to as a *NaTech* (Phillips et al 2012).

Before delving into FEMA's preparedness cycle (FEMA 2010), in the following sections, let's briefly look at FEMA's phased approach to emergency management. This is often referred to as the *disaster or emergency management life cycle*. Since disasters are not a singular point-in-time event, they rather have a life cycle of occurrence. This cycle triggers a phased approach to emergency management. The phases reflect the thought and physical processes used to address and handle a disaster. The phases are *prevention*, *protection*, *mitigation*, *response*, and *recovery*. These will be discussed throughout this chapter and will be woven into the discussion on FEMA's preparedness cycle.

Planning

FEMA defines emergency planning as a continual cycle that includes planning, training, and exercises, and continual revisions that take place throughout the five phases of the emergency preparedness cycle. The purpose of the planning process is to develop and maintain an evolving emergency operations plan, which outlines how potential hazards are responded to. Keys to establishing a proper emergency planning process include developing an effective and properly staffed planning team, understanding the issues that are to be faced, determining what is needed to have an effective and scalable response, and the preparation, review, and implementation of the plan.

This ongoing process, which starts when considering how to prepare for and respond to a disaster or incident, continues and is retooled during recovery. From a security perspective, the hazards and threats of an incident or event, plus potential sabotage, should be considered. The planning team should consider and coordinate with, among other potential partners, TSA, FRA, United States Coast Guard (USCG), and transportation providers, when establishing the planning team.

It has been suggested that this process was not fully in place for Hurricane Katrina. But after the railroads in New Orleans experienced significant problems, to include stranded and damaged hazmat cars, and perceived lack of security, the railroads, TSA, and FRA became active contributors to the lessons learned evaluation and the subsequent planning process.

When Hurricane Rita struck a month later, the problem was exacerbated somewhat because many of the rail cars that were evacuated from New Orleans were now in Houston congesting their yards and tracks. This experience also served to further hasten the involvement of the railroads, TSA, and FRA in the planning process. It was quickly realized that the hazmat cars, when unattended, stranded, and/or damaged, or unable to be moved because of congestion in the yards and on the mainlines or due to track damage, made attractive targets for vandalism or terrorism.

Preparation

The National Incident Management System (NMS) identifies a continual preparedness cycle, as shown in Figure 10.2, to prevent, respond to, recover from, and mitigate incidents and disasters. Preparation, in this sense, includes organizing measured responses and resources consistent with the need to effectively respond to incidents and disasters.

Training and conducting exercises is an important and continuous task and is conducted throughout the cycle. Planning is an effective tool that continues into the preparation phase. Training and exercises in particular are helpful, as they test the response plan and ability of partners to work together and understand each other's capabilities. The preparation phase is when available resources are matched with needs and mutual aid, and contingency plans are developed to address and supplement the needs and shortfalls. This includes, but is not limited to, first responders, supplies, evacuation and sheltering, support services and personnel, and equipment. Regional planning helps assure that the resources are available and in locations that are not potentially impacted by a disaster or incident. For example, emergency resources for Houston are in Austin, Dallas, and San Antonio for Gulf Coast hurricanes.

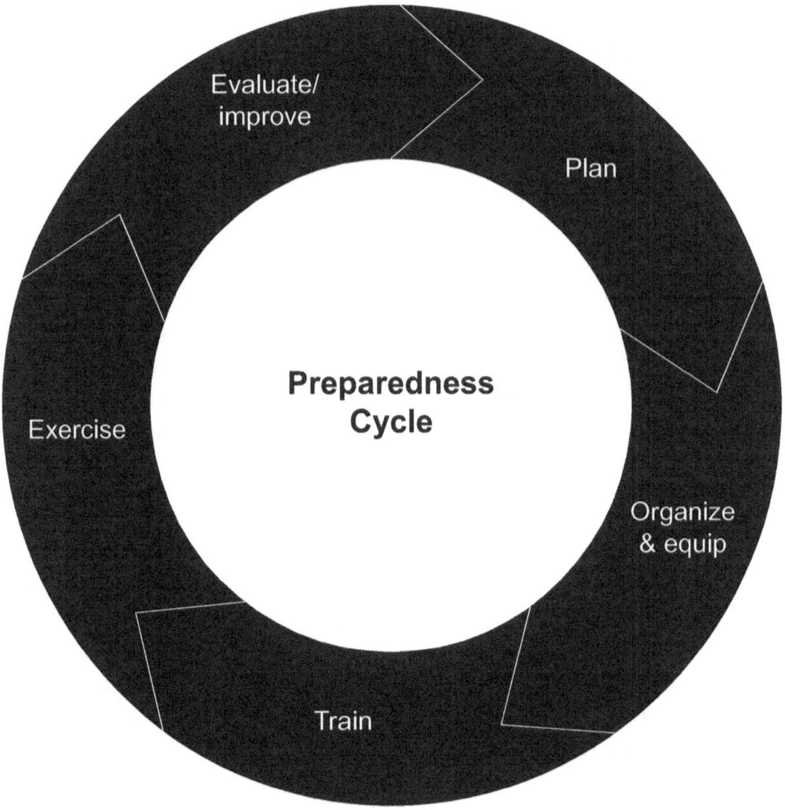

Figure 10.2 FEMA preparedness cycle

Source: Federal Emergency Management Agency

Preparation is not just during the 120 hours before a hurricane's landfall or when a blizzard is to occur. It is begun often as planning months or years before and includes the identification and coordination of resources to match the needs and shortfalls. Much of this occurs among and between FEMA, other involved federal agencies, state and local emergency agencies, first responders, and law enforcement. The railroads, transit systems, and other transportation providers are also involved, as they often are used to evacuate people, remove debris, and haul in material to effect repairs. Also, and in the case of freight railroads and hazmat cars, they are involved to eliminate the inherent dangers associated with unprotected rail cars that contain hazardous material that could compound a natural disaster.

As a part of the preparedness phase, annual and periodic planning conferences and exercises are conducted to, among other things, identify gaps and needs, exercise the emergency response plans, and test partnerships and communications. And by doing so throughout the year, problems can be identified and resolved before a disaster or incident occurs. This also allows response to be properly scaled to address the magnitude, type, and geographic conditions of the disaster or incident.

From a security standpoint, the transportation providers and applicable government agencies are able to identify and address potential threats and vulnerabilities during the preparedness phase. This is important, especially when all are "hunkered down" for a forecasted natural disaster and during the initial response. For vandalism or a terrorist incident, a threat and vulnerability assessment needs to be prepared and be both current and flexible. In this case, there is little or no warning, hence planning and preparation must be an ongoing process that is flexible and executable at a moment's notice.

An example where preparedness helped was when Hurricane Rita approached the Gulf Coast in 2005. An unclassified DHS message was passed down to the TSA region that stated "chatter" had been noticed indicating a desire to attack hazmat rail cars along the Gulf Coast during or immediately after landfall. Preparedness activities post-Hurricane Katrina, even though it had only been several weeks, allowed for the expeditious dissemination of the warning to the railroads along the Gulf Coast. Appropriate action was taken by the railroads, and no event occurred. This notification process could be used for any type of disaster recognizing that warning times differ and may not afford a complete dissemination of information and taking of appropriate action.

Railroads during this stage evacuate rail cars, particularly those containing hazmat, or protect them in place. The protection includes not only damage from a storm but that for vandalism or terrorism, in conjunction with the storm or as an event by itself. This can be difficult because of the need to protect life and limb, and it can require sacrifice and/or electronic means of surveillance and monitoring (i.e. camera and intrusion detection). If there is a terrorist incident, there is generally no warning, so the preparation must be in terms of heightened awareness via established plans and procedures on what to do given short or no notice at all. This is the challenge, and planning is essential to ensure that the potential hazards and associated vulnerabilities are identified and that contingency planning and resulting readiness are in place.

During/landfall

During a hurricane's landfall or during an incident, attention is focused on survival and protection of property. First responders are hunkered down and ready to go, while railroad operating and maintenance personnel, to include railroad police, are planning the inspection and repair activities and preparing for triggering the operating contingencies developed. For hurricanes, storms, and events with warning, the prepositioning of personnel, equipment, and material is essential and accomplished during planning and preparation. This is often at locations that are not in the path of the storm. For example, the BNSF and UP railroads staged crews and maintenance personnel in San Antonio, Austin, and Dallas / Fort Worth in anticipation of landfall for Hurricanes Rita and Ike. Similar staging was used for Hurricane Katrina, but not to the extent developed from lessons learned.

Snow storms pose a somewhat different problem to railroads. Once power is restored (either permanently or temporarily), the tracks are cleared, and the bridges are inspected, railroad operations can be restored to a somewhat normal level. In 2015, the Northeast experienced frequent and heavy snow storms endearingly referred to *Snowmageddon 2015*.[2] Snow storms often have similar problems as hurricanes, but it takes longer to resume operations. Some of the reasons include snow plowing and clearing the snow from the tracks, stations, and rail yards; adverse weather conditions that may affect power restoration; and the effects of the cold *and* snow on the locomotives and rail cars. This exposes the railroads to the additional risk of vandalism and terrorism because of the longer service outage. Note that blizzards, as was evidenced in 2015, can be Nor'easters or snow hurricanes.

Hurricane planning and lessons learned have shown that local and state law enforcement restricts access to affected areas to prevent, among other things, looting, risk of injury, and further and secondary damage. This was the result of lessons learned from Hurricane Katrina. Experience with Gulf Coast hurricanes has shown that TSA helped the railroads gain access to affected areas to assess the damage and determine repairs needed. This is important for many reasons. First it enabled the railroads to inventory and assess the condition of the hazmat rail cars not evacuated. Then and equally as important was to get the railroad up and running as soon as possible so that it could assist in debris removal and bring in construction equipment and material to aid in the rebuilding – and, of course, repair and/or evacuate the hazmat rail cars that may have been damaged. This process is finalized during the preparedness phase, but the implementation strategy is reviewed during landfall so that the response is quick, effective, and seamless as soon as the storm or incident is over to continue to the next phase of the emergency management plan.

This phase is where the railroad and public in general are most vulnerable. Attention is focused on survival and protection of assets. Vandals may have some values with regard to their well-being, while terrorists do not. Sabotaging rail cars, equipment, and infrastructure, such as signals and bridges, could go unnoticed. The damage may not be noticed until recovery begins or an incident occurs because of sabotage that went unnoticed. This sets the stage for innovative technology to protect the rail cars, equipment, and infrastructure when "hunkering down," coupled with procedures that prevent access or an incursion (i.e. fencing, CCTV, and burying hazmat rail cars deep in the yard to prevent easy access). The railroads must be diligent in inspecting every asset during the response and recovery phases and before service is restored.

"Hunkering down" and preparing for response and recovery is relatively straightforward when there is sufficient advance warning. For tornados, earthquakes, and terror attacks, there is minimal or no notice. There is essentially no chance to "hunker down" and use that time to refine the emergency operations plan based on the event at hand and modify the response and recovery activities accordingly. This makes it much more important that the planning and preparations take an *all hazards approach* and not just one that addresses the likely disasters and incidents.

Response

Once the storm passes or incident wanes, railroad personnel, often accompanied by first responders and law enforcement, inspect the track, right-of-way, bridges and structures, signals and communications systems, and facilities to determine the extent of damage, if any, and impact on operations. In today's environment, this includes looking for signs of sabotage or tampering with equipment and infrastructure. Operating and mechanical personnel inspect the locomotives and rail cars for damage, spills, and leakage. Spills and leakage of hazmat could pose a problem and may cause delays in fully responding and starting the recovery.

After Hurricane Katrina, CSX had many hazmat rail cars stranded at its Gentilly Yard in New Orleans because of the flooding. Personnel had to respond to determine if there were any hazmat releases before the inspection could be completed and recovery efforts could begin (Gordon and Secrest 2006). After Hurricane Ike, a potential terrorist issue at the BNSF causeway and bridge to Galveston needed to be resolved before the inspection was completed and the recovery plan initiated. This was important, as the railroad was to play a key role in hauling debris out of Galveston and construction material in – the railroads can often be a critical resource for affecting a timely recovery from any disaster, whether they be natural or manmade.

The response and subsequent inspection for rail transit systems and passenger railroads is basically the same for a freight railroad. The difference is that stations and terminals are involved and, for rail transit systems, Amtrak, and some commuter rail operations, third rail or overhead catenary and the associated power distribution systems must be inspected. Amtrak and commuter railroads are often in shared corridors or utilize the same tracks as freight railroads, so there may be a shared response and inspection covering both operations. It is important that the inspection look carefully for signs of tampering, which may not be apparent, or placement of IEDs. Also, structural stability could be compromised by the incident or storm and could be exploited.

These principles can also be used when responding to a terrorist attack or natural disaster, other than a hurricane. For a terrorist attack, the inspection should consider that the attack may be the first of a series of coordinated attacks focused on multiple sites, first responders, detour routes, and impacting passenger trains in the case of freight rail. For passenger rail and in addition to that for freight rail, consideration must be given to a secondary attack on evacuating passengers and land uses adjacent to large stations and terminals. Therefore, care needs to also be taken to protect the first responders, individuals inspecting the damage and evacuees.

Recovery

Once the damage is assessed during the response phase and early stages of recovery, it will become apparent as to what is needed to make the necessary repairs to resume operations, as quickly as possible. This is an important step in returning to normalcy, which includes regular "eyes on" the operations, infrastructure/

facilities and equipment. The lack of "eyes on" could increase the risk of sabotage and vandalism. However, the desire to resume normalcy should not forsake a thorough inspection and assurance that the facilities and structures have not been compromised. Also, temporary repairs should be made so as not to lend itself to easy tampering.

A railroad or rail transit system's continuity of operations plan (COOP) should address expediency and the potential issues arising from the incident or storm. A key thought is that, once the system is operating as close to normal as possible, the built-in security and awareness procedures resume normal operations. Also, public confidence will be uplifted when operations approach normal. This includes safety for rail passengers and ability to monitor and safely handle and track hazmat rail cars.

During recovery and especially if design is involved, consideration should be given to improving the security at facilities (i.e. rail yards) and structures (bridges and tunnels). Infrastructure protection and hardening, system redundancy, and CPTED-focused designs (Fennelly and Crowe 2013) are ways in which to accomplish this. When replacing structures and facilities in kind, using more resilient material should be considered, and systems should reflect state of the art technology.

Agencies and stakeholders

Federal government agencies

Linking emergency response with security is a multilevel and multiagency function. At the federal level, FEMA is the lead emergency response agency for DHS. For storms and natural disaster, FEMA works closely with other DHS component agencies, to include TSA and the USCG, the U.S. Army Corps of Engineers, U.S. Department of Transportation agencies, and Environmental Protection Agency (EPA). Each has its own function in NIMS and the incident command structure. Based on the magnitude of the disaster, a joint field office (JFO) may be set up. The security component of emergency management and in particular with regard to aviation and surface transportation is the responsibility of TSA. TSA collaborates with the USCG at ports and with FEMA in the operation of the JFO and "down range". TSA works with the railroads and rail transit systems to ensure that their facilities and equipment are protected. In the case of freight railroads and, especially hazmat rail cars, evacuation or securing them in place is paramount. Pre-planning and exercising is helpful in preparing for a natural disaster. And in the case of hurricanes, there is time to implement and "tweak" the preparedness plans and adapt for any changes.

In the event of a terrorist attack, pre-planning, training, and exercises are crucial. This is because there can be little or no warning of a disaster. Based on lessons learned and forward thinking, DHS has a series of training and exercise programs that allows the railroads and transit systems and their guiding and regulating government agencies to test their plans and procedures involving a disaster whether manmade or natural. On the surface transportation side, there is the Intermodal Security Training and Exercise Program (I-STEP). On the maritime side, there is the Area Maritime Security Training and Exercise Program (AMSTEP). Surface transportation and railroads are an integral part of both exercise programs. The exercises "game" terrorist events, natural disasters, or in combination. This allows the emergency planners the opportunity to build upon "lessons learned" without a disaster. Further, the railroads can better understand the risks and possible occurrences that could happen during and after a disaster, especially when hazmat is involved.

As part of and annexed to the National Response Framework (NRF) are the Emergency Support Functions (ESF). According to FEMA the ESFs are structured to provide coordinated support to the federal response to an incident. The ESFs are categorized to provide coordinated and needed support, based on designated functional areas and tasks. When looking at railroads, ESF #1, Transportation, and ESF #3, Public Works and Engineering, are the ones with which the railroads are most likely to be involved. As experienced by past hurricane involvement, the railroads could provide transportation for mass evacuations and haul away debris and haul in construction material. From a security perspective, the railroads, to provide the support needed, must be "up and running" as quickly as possible and would have to recover in an expeditious manner. Inherent in the quick recovery is a rapid return to relative normalcy with associated security and monitoring in place. This will benefit and lower the risk of hazmat rail cars being compromised post incident when the railroad is busy cleaning up and restoring operations.

State, regional, and local government agencies

State, regional, and local governments are the "boots on the ground" dealing with disasters, to include security related events. They are the "first responders" to a disaster and work in consonance with the federal agencies that regulate processes and operations, and they are intended to render assistance and protect the general public.

Fusion centers, such as the El Paso Intelligence Center (EPIC), Boston Regional Intelligence Center (BRIC), and California State Threat Assessment Center are routinely set in major metropolitan areas, regionally, or statewide to coordinate emergency response, share intelligence, and provide a cooperative environment for all involved agencies to partner year round. This provides a foundation and backing should an emergency arise requiring incident command. Further, the fusion centers work with the railroads' operations centers in monitoring operations that could be impacted by or cause a disaster in the region. This includes the handling and tracking of hazmat rail cars and monitoring weather and other conditions that could cause a disaster. An example of this is the Union Pacific Railroad's Harriman Dispatch Center in Omaha, NE, where rail lines in California are monitored for seismic activity that could be a precursor of an earthquake. Similar monitoring on other railroads includes that for flooding, slope stability, groundwater, etc.

As stated previously, hazmat trains are monitored by fusion centers in coordination and cooperation with the railroad operations centers to ensure that disasters do not occur and the railroads' operations are not compromised. The sharing of hazmat routing and commodity information is important among the proper entities, to include first responders, based on a need to know. The wide dissemination of this information has come to light in recent years with several states wanting to release the routing and commodity information to the general public. This creates challenges for and risks to the railroads, shippers, and receivers of the hazmat, and state, local, and regional first responders and those entrusted with protecting the public from a potential disaster. If the information were to be disseminated to the general public, terrorists could use the information to plan and possibly carry out an attack that could have dire consequences. This is an important consideration when planning for and mitigating the impacts of a disaster and preparing for response.

Incident command in an emergency is set up in accordance with FEMA guidelines and its leadership. It can be collocated with fusion centers or at JFOs set up for the incident or at predesignated state, regional, or local locations. Based on the type of incident and extent of railroad involvement, the railroads, usually through their operations centers, could be located with incident command or will be in direct communication with it. State-level agencies that deal with railroad operations and safety include the Massachusetts Department of Transportation Rail and Transit Division, and Texas Department of Transportation Rail Division. They, in turn, closely coordinate with the federal agencies that address railroad safety. This partnership is only as good as the partnerships built with the railroads. For example, the FBI Joint Terrorism Task Forces (JTTF) at the national and metropolitan area levels often have law enforcement representatives from the railroads and state, regional, and local jurisdictions. The JTTF works with its member agencies and organizations in sharing intelligence that can not only thwart an event but aid in preparedness planning and developing a COOP for a disaster regardless of type.

Stakeholders

Organizations that feel the direct impact of a disaster on its operations and facilities are the stakeholders. They are the companies that must recover from a disaster and have the personnel, equipment, and financial resources to do so. Also, they have the arrangements with contractors to assist in debris removal and reconstruction efforts. In the railroad community, these are the freight railroads, holding or operating companies, and transportation authorities that have commuter rail operations. Tourist and scenic railroads are part of the "population" but are not involved in transporting hazmat or any commodity that would be of interest to a terrorist. However, they could be a target if mass casualties and sites of iconic value are the intended objective of the terrorist. Examples are the Mount Washington Cog Railway[3] in New Hampshire and Alaska Railroad.[4]

In addition to the large Class I[5] railroads, you have shortline, switching and terminal, and regional railroads. The Class I railroads have the financial resources and personnel to plan and prepare for, respond to, and recover from a disaster. This effort is often led by the police or resource protection departments in conjunction with the maintenance and engineering, mechanical, and transportation departments. All of these departments have priorities and directed activities with regard to protection, response, and recovery. For example, the maintenance and engineering folks are concerned with protecting the infrastructure, hardening it, and/or providing redundancy and resilience – the latter so as to quickly repair the infrastructure to all for a speedy recovery of operations.

Because of the lack of resources associated with lower operating revenues, the shortline, regional, and switching and terminal railroads have less resources and finances to monitor, respond to, and recover from a disaster. Many of these railroads are state owned or operate on state-owned rights-of-way (i.e. the Alaska Railroad is a state corporation) or owned by a parent organization, as one of many railroads (i.e. Genesee & Wyoming owning about 120 railroads worldwide). The resource problem in this instance is less than with the independent smaller railroads. One benefit, albeit small, is that the shortline, regional, and switching and terminal railroads are smaller in size and have less territory and facilities to monitor and protect.

Throughout these phases, the security of the railroad's infrastructure and equipment must be planned for and maintained, particularly during the period immediately after a disaster occurs. Therefore and to this end, the railroads have developed a working relationship with both the FRA and TSA beyond that required by regulation with regard to its operations and infrastructure that supports it. This partnered approach is more effective than regulations. The railroads also work with FEMA and state departments of emergency management (DEM) with regard to preparedness planning,[6] training and exercises, and lessons learned as they help in the planning for future disasters. Working with law enforcement at all levels on matters of intelligence sharing is important in that it could be instrumental in deterring and detecting a potential threat that could take advantage of the protective, "hunker down" posture taken during a storm.

Redundancy and resilience or ability to rebound from a disaster is important to security (Phillips et al 2012). This is because a fully operational railroad has all of its physical and operational security components in place. But by providing system redundancy, a railroad may be lessening its resiliency. A railroad that has a redundant dispatching capability could feasibly resume operations sooner, but it may be at the expense, both operationally and financially, of making other system components, such as track, structures, and signals, as resilient as possible. It is believed that Hurricane Katrina contributed to the Kansas City Southern, relocating its primary dispatch facilities from Shreveport, LA, to Kansas City, MO. Now, a Gulf Coast hurricane will have far less of a potential to impact both the dispatching capabilities and track and structures simultaneously.

In the early days of railroading, it was possible to detour a train around a disaster. This is because there were significantly more and redundant route miles and supporting structures then than there are now. Given this fact today and the need to resume operations as soon as feasibly possible, a balance of redundant and resilient systems is important. Therefore, the railroads should carefully weigh and balance system redundancy and resilience, when planning for a disaster, and know the type and intensity of a potential disaster. This is important when addressing the impact of disasters that have advance warning, such as a hurricane, or no or minimal warning such as a tornado, earthquake, or terrorist attack. And what about a NaTech?

What can the railroads do?

Based on the above combined with the principles of emergency management and railroad operations and infrastructure, let's look at what railroads can and need to do to maintain a positive security posture in the face of a disaster.

The following are concepts with regard to a railroad's establishing and maintaining a safe and secure environment before, during, and after a disaster. There are many ways to accomplish this endeavor and documents used to ensure this. Emergency management is not a precise science and is based on, among other things, historical data, lessons learned, research, and physical and social sciences. Therefore, this section will address in general terms what approach a railroad should consider and a general description of the documents that could help achieve this.

Planning stage

During the planning stage, the railroad should work with its internal departments and shippers and receivers to determine the needs, both operationally and physically, to provide for reconstituting service as soon as feasibly possible after a disaster. This is accomplished via the development of a disaster/incident response plan and COOP. The documents should address, among other things, infrastructure protection and hardening, backup systems, what must be evacuated (such as hazmat cars that could pose a serious problem should a release of material result), and what can be protected in place. You do not want to have to determine what must be protected and what must be evacuated as a disaster becomes apparent or approaches. What is to be done is identified during planning and "spun up" during the preparedness phase. For unforeseen events, such as an earthquake or terrorist attack, prior planning is very important.

Railroads should carefully and methodically plan out and document their course(s) of action. The approach should be *all hazards* and document the general and disaster-specific recovery methods (in annexes). It should address, among other things, the physical plant, operational contingencies, recovery operations, interoperability of communications, and safety of personnel. Identifying what is needed and availability, and where to pre-position the resources, equipment, and material is important. This is because, if left to the preparation stage, it could be ineffective and untimely. Once the plans are developed, training the personnel on and exercising the plans is important.

Preparation stage

During the preparation stage, the railroad should follow the steps identified in the planning documents and as trained and exercised. This includes disasters with little or no advanced warning. The resources, equipment, and material needed to recover should be pre-positioned at this time and infrastructure protected and hardened, as appropriate. This includes the pre-positioning of locomotives so that they are not in jeopardy and available for recovery and reconstituted operations. Also, it is important that the T/PIH and hazmat and high value, by commodity shipped (i.e. military shipments) and dollar value, should be evacuated or protected in place – the former to ensure that they are not compromised and cause a combined disaster. The importance of this was evident when a train carrying military equipment was stranded on a shortline in Southeast Texas post–Hurricane Rita, exposing it to risk. For disasters with little or no warning, especially a terrorist event, the operational posture of the railroad should be such that trains can be diverted and protected in short order.

Redundant systems should be inspected to ensure that they are operating and ready, and equipment, especially hazmat rail cars, should be evacuated or protected in place based on timing and notice/forecast and line capacity. This is relatively easy for disasters with advanced warning, but, for disasters where there is little or no warning, it is not as easy. Therefore, advanced preparation, to include determining hardening solutions and measures when protecting in place, is important. This includes, but is not limited to, the identification of infrastructure to be protected and hardened so as to minimize vulnerabilities based on known, anticipated, and combined threats. Problems identified during this stage should be used as lessons learned in future disaster planning.

Railroad line capacity became a problem with Hurricanes Katrina and Rita because they were only 30 days apart. What was evacuated to Houston from Hurricane Katrina "clogged" the rail lines and yards, making evacuation for Hurricane Rita difficult. For Super-Storm Sandy, the intermixing of freight, commuter, and Amtrak trains made evacuation and protecting operations and equipment difficult. In 2015 and for the constant heavy snow storm in the Northeast, clearing the mainlines, yards, and stations made pre-positioning and protecting equipment difficult.

During and landfall

During a disaster, to include landfall of a hurricane, the railroad must, to the extent possible, monitor critical infrastructure and equipment that has not been evacuated. This can be accomplished by the placement and protection of key personnel in the "impact zone" and having others readily available in a safe haven close by to "resurface" immediately after the disaster. When there is little or no advanced warning of an incident or disaster, knowing the potential threats can help in having "eyes on the target", even during the incident or event. But the railroad or other entity should use remote monitoring as much as feasibly possible. Regardless, people should not be placed in harm's way unless absolutely necessary and a credible threat to the railroad and its facilities exist. Just knowing of a potential threat should not compromise the safety of personnel.

Given the above, effective communications are essential. Recognizing that normal communications could be affected by power outages, communication lines that are down, and cell tower damage, alternate communications (such as satellite radio and predetermined lines of alternate communications) should be prearranged. Communication interoperability is also important, as your communication system may be down, but a partner's system, such as the local fire department, may be up and running. They, in turn, could provide you with "bandwidth" or relay important messages until your system is back up and running. This is important when communicating within your "stovepipe" and across the spectrum of the others involved: the JFOs, fusion centers, or emergency operations centers being paramount.

Response stage

After the incident, disaster, landfall, etc., the railroad will respond with its resources in waves. The first will be by those in close proximity to the event and/or "hunkered down" and then followed by additional personnel, to include technical experts, such as bridge inspectors, to assess the damage and determine the extent of repairs needed. Based on the location of the damage, first responders, to include law enforcement, EMS and fire departments may be in the mix of initial personnel on the scene. It is important that all persons and organizations responding do so in a coordinated manner and effective communications are established. FEMA's incident command system will assure that this is accomplished and, if there is sufficient warning such as with a hurricane, and the anticipated magnitude of the event or incident is large enough, incident command would be instituted during the preparedness phase.

When there is little or no warning and at locations where there are many people, such as a multimodal train station, where there is bus, transit, and rail, response is based on careful planning among all partners using the facility. A preplanned incident command system is essential in this case to avoid mass confusion and chaos during the response phase of a disaster. Preplanning, and training and exercising the scenarios, will be instrumental in having effective and measured response procedures in place when needed.

You can have a well-planned response plan and effective and interoperable communications, but unless you have coordinated access to the impacted area, your response could be rendered ineffective. From hurricanes, the 9/11 attacks, and other disasters, safety and security warrants cordon lines that delineate restricted access areas that are managed and controlled by law enforcement, etc. Therefore and during the planning phase, access control should be addressed and methods of gaining access should be developed. This is often accomplished by vetting key personnel and those responsible for safety and security of the restricted areas. By doing this, the railroad will be able to access their facilities impacted by the incident or event and assess damage and safety concerns with minimal effort. Post Hurricanes Katrina and Rita, processes were instituted to ensure escorts were provided to access restricted areas and "badging" to allow individuals to access restricted areas unescorted. The "badging" would be similar to the joint TSA/USCG Transportation Worker Identification Credential (TWIC).

Recovery operations

At this time, incident command should be in place and working with all entities in their recovery efforts, to include the railroads. Full recovery early on is not or should not be expected, so a logical and phased recovery plan should be enacted. The railroad and its contractors should be in communication with incident command to ensure that their efforts are in consonance with the other stakeholders and coordinated with the ESF lead agency. That would be the U.S. Department of Transportation for Emergency Support Function #1 and U.S. Army Corps of Engineers for Emergency Support Function #3.[7] For rail security measures, TSA is the point of contact within ESF #1. Infrastructure and facility repairs should be underway and prioritized based on operational need.

The focus of the railroad's activities should be to prioritize the restoration of service and open up rail lines to assist the recovery efforts, as well as evacuating hazmat cars to minimize risk. This includes hauling away debris and bringing in construction material to assist with the recovery efforts. The focus should also be on bringing in locomotives to have the "power" available to resume operations and move the hazmat cars out of harm's way. If hazmat rail cars are left unattended for too long or if they are damaged, they could be susceptible to tampering. Also, coordinating with shippers and receivers to move rail cars in and out, as needed, will occur during this stage and, perhaps, minimize their risk of compromise.

During this stage, the railroads, agencies, and other stakeholders should, based on what is happening, be looking at, among other things:

- Mitigation planning and preparing for events and incidents better, based on lessons learned.
- The effectiveness of hunkering down and protecting and monitoring their assets and how to improve upon what has been done.
- Responding after landfall or after the disaster occurred, and ability to access or provide access to restricted areas.
- How effective the communications were and especially interoperability.
- The effectiveness of the recovery efforts, and were they focused enough so that operations were restored in a logical and timely manner.

The above should be addressed in a "hot wash" and discussed as lessons learned. The short falls or gaps identified should be incorporated in revisions to the planning documents, and then trained and exercised.

Continual planning and preparation

The FEMA preparedness cycle promotes effective planning, organizing and equipping, training and exercising, and evaluating and improving preparing for a disaster. By doing this, response will be more timely, "on point", and organized. Lessons learned provide the basis for the railroads, stakeholders (i.e.

shippers and passengers), or other involved entities (regional transportation agencies, state departments of transportation, and state and local emergency management agencies) to evaluate and improve on response operations and procedures resulting from what was just experienced. This evaluation can be used in developing a new planning document that fully addresses the results of the improvements recommended from lessons learned. Next, the plans for improvements, along with the other components in the plan, are organized and coordinated with regard to resource and equipment availability and scalability, and commitment to provide them. Then, personnel and organizations should be trained on the processes highlighting what is revised per the lessons learned and, after suitable and focused training, exercised. It is hoped that this can be accomplished before next hurricane or tornado season, disaster occurs, threat arises, etc.

The above should be well documented in proper revisions to the appropriate response and continued operations plans. Throughout the process, effective communication and coordination should be encouraged. Without this, good planning could be rendered ineffective. The result of proper planning, preparation, and execution of a response or other disaster-related plan is the ability to maintain security during the incident to the extent possible, maximize the ability to respond quickly, and recover and restore operations as soon as feasibly possible given often conflicting priorities and what may appear to be insurmountable constraints. Attending and moving rail cars, particularly those containing hazmat and T/PIH, pose less of a security risk than those that are stranded.

Summary thoughts

Much of this chapter has dealt with the emergency management phases of planning, preparedness, response, and recovery. The railroads, perhaps more so than any other mode of transportation hold a unique position in the U.S. economy given that (1) most of the trackage is privately owned, (2) they participate in very proactive industry associations, namely the AAR and ASLRRA, (3) they have enjoyed a long legacy of mutual assistance in the face of disasters, (4) the impact from the loss of rail service, even if only regional, can be significant for the U.S. economy, and (5) the railroads can be a valuable asset in aiding with any post-disaster recovery.

This chapter opened using an impending hurricane as a disaster scenario and highlighted the range of initiatives that had to be considered when making any kind of preparations. Nevertheless, every disaster is different whether one considers its overall scope, the specific types of railroad assets that will be potentially impacted, and the nature and quantity of freight and/or passengers that will be affected. The planning models that have been discussed use a commonly accepted FEMA template, but executing this with railroad organizations may well represent the best that a public-private partnership can possibly offer.

Notes

1 When discussing an all hazards approach, it must be understood that each disaster, whether natural or manmade, will have similar characteristics warranting a uniform or common set of management strategies and procedures. The differences are typically addressed in disaster specific appendices.
2 *Snowmaggedon*, a 2011 made-for-TV movie written by Rudy Thauberger, is the term used by the press and in particular the *New York Post* to describe the repeated and heavy snow storms in 2015 in the Northeast.
3 While this is a useful example here, tourist railroads were excluded from the scope of this book because of their usually modest geographic footprint and minimal economic impact if service was to be disrupted.
4 The Alaska Railroad is a 500-mile Class II (as originally defined by the former Interstate Commerce Commission based on operating revenues) or Regional Railroad (as designated by the Association of American Railroads for most Class II railroads) that provides freight services as well as significant and seasonal scenic and tourist passenger service from Seward to Fairbanks, Alaska.
5 According to the Association of American Railroads, Class I railroads concentrate largely on long-haul, high-density intercity traffic – operating in 44 states and the District of Columbia. They are regulated by the Surface Transportation Board of the U.S. Department of Transportation and have annual revenue exceeding $453 million, and account for 69 percent of the industry's mileage, 90 percent of its employees, and 94 percent of its freight revenue.
6 Although planning is part of preparedness, the term *preparedness planning* is often used to describe the relationship between planning and preparedness.
7 *Emergency Support Functions* (ESFs) are defined by FEMA as key elements contributing to the success of the *Incident Command System* (ICS).

Bibliography

Federal Emergency Management Agency. (2008). *Incident command system training.*

Federal Emergency Management Agency. (2010). *Developing and Maintaining Emergency Operations Plans Comprehensive Preparedness Guide (CPG) 101*, Version 2.0

Federal Transit Administration. (2003). *The public transportation system security and emergency preparedness guide.*

Fennelly, L., and Crowe, T. (2013). *Crime prevention through environmental design* (3rd ed.). Portsmouth, NH: Butterworth-Heinemann.

Gordon, G., and Secrest, C. (2006). *Railroad security efforts post Hurricanes Katrina and Rita.* Paper and Presentation, American Railway Engineering Maintenance of Way Conference.

Phillips, B., Neal, D., and Webb, G. (2012). *Introduction to emergency management.* Boca Raton, FL: CRC Press.

11 Protecting rail infrastructure

Where do we go from here?

Lacking any empirical evidence other than rail attacks in other nations, public policy and industry responses remain rooted in speculation as to the big questions regarding rail security in the U.S. Is the security risk to the nation's rails real? If so, who is the likely perpetrator, and what are the actions most likely to form the major threats? How vulnerable is the nation's rail network, and what can be done to proactively plan for both prevention and response? Should rail security be seen as a major component of homeland security, and, if so, what resources and actions are needed by policy makers and rail operators?

It is our assessment that the rail industry in the U.S. is at the epicenter of a perfect storm: the origins for which can be found in its ownership distributed across a patchwork of interdependent private firms, public agencies, quasi-public authorities, and the National Rail Passenger Corporation (Amtrak). The U.S. rail network operates in diverse settings that include densely populated urban areas, such as those found on either coast, and sparsely populated, desolate terrain that includes mountains and deserts. The conventional means of policing, such as infrastructure protection and handling trespassers, has always been challenging, but some events of recent years suggest that such may be even more difficult today.

Then there is the widely held public perception that railroads are an antiquated and obsolete mode of transportation. The public at-large, with perhaps the exception of the regular users of rail passenger service (e.g. commuters), are substantially oblivious to the rail networks unless they happen to be inconvenienced having to wait for a passing train at a grade crossing or live adjacent to a busy rail line or be disturbed at night as they pass by. Few realize the volume of freight – whether as basic raw materials such as iron ore, coal, or salt or as finished goods moving in ocean containers and highway trailers – that constitutes a major and growing portion of the U.S. economy. Too few recognize from an investment standpoint that the railroads have once again become growth stocks worth owning. In recent years, the freight rail industry has stepped up its efforts to publicize its strengths, but the gap in public perception remains.

The railroad industry has been transformed in several key ways. First, there is the dramatic increase in intermodal transportation where a very large percentage of containers and highway trailers travel some portion of their route between origin and destination over the rails. Second, there is the substantial reduction in the number of corporations within the industry. Most notable has been the decline in the number of Class I railroads from several dozen half a century ago to only seven if one is to include the U.S. operation of the two major Canadian railroads, both of which by virtue of size are Class I's. Consolidation is also affecting shortline and regional railroads. While there is a growing number of passenger services in the form of new commuter routes and expanded corridor operations, Amtrak remains the nation's only long-distance passenger operator and is stuck in a never-ending fight with Congress to maintain adequate funding and operational control. Compared to the situation in 1971 before the creation of Amtrak, freight and passenger operations now often share rights-of-way and often tracks but not ownership, complicating the relationships needed to ensure rail security.

Public perceptions

The negative public perception of railroads in general, meaning both passenger and freight, is subsequently echoed by public officials at all levels, who then tend to minimize the importance of the rail network and what the threats to it might be. By contrast, the concern for air transportation following the attacks of 9/11 has resulted in myriad new regulations and spawned the creation of the Transportation Security Administration, a federal agency that devotes most of its resources to screening airline freight and passengers with a disproportionately low attention to rail and other modes of surface transportation. The regulations protecting rail infrastructure and operations afford but limited penalties focusing on those caught trespassing. However, recent events are seeing more proposed regulations focusing on operational security, to include training of personnel, chain of custody of certain categories of hazmat, hazmat routing, and potentially new vulnerability assessments and security plans. Moreover, the resources for policing that infrastructure have not expanded appreciably to coincide with the increase in threats.

We posited in an earlier chapter that passenger railroads make better terrorist targets than freight trains because of the panic that ensues and the publicity factor accruing to the terrorist organizations from successful attacks. Grisly images of dead and injured passengers raise fears more than pictures of derailed freight cars. Moreover, terminals and stations make for inviting targets when the choice is commuter trains, but the trains themselves make for better targets when considering intercity rail travel. These statements, however, must be seen in perspective, as any approach that strictly segments the railroad network into discrete units by type of freight or passenger performs somewhat of a disservice to the reader because of the joint use of infrastructure, in particular the sharing of rights-of-way and other infrastructure. With the common use of rights-of-way by both freight and passenger operations, the ability to address threats as discrete by type of operation becomes an unrealistic approach to any thinking on the issue of protection. The balance of this chapter will instead divide infrastructure protection into the subtopics of human assets, technology applications, public policy initiatives, and operational security. Practicing more of the same based on the past will without anticipating new threats and vulnerabilities eventually lead to something akin to the military parallel of preparing to fight the last war instead of the next one.

Complexity

In our first chapter introducing the topic of protecting railroad infrastructure, we made a case that the difficulty of protecting rail infrastructure was to be found in the complexity and expansive nature of the U.S. rail system. The railroad network is really an amalgam of interconnected firms and agencies that own or otherwise control their own rights-of-way, equipment, information systems, and terminals or stations. These organizations also have different missions and operations depending on whether they are freight railroads, intercity passenger carriers, or commuter operations. In some cases they may run over lines that they do not own and may operate locomotives and rolling stock that belong to others. Their rights-of-way may be largely urban, as in the case of the commuter lines, or have hundreds of miles of isolated track running through desolate sparsely populated territory.

Their cargoes are the preferred targets, whether these are human or material, yet even when the target is material, the ultimate objective remains doing harm to humans and/or causing economic disruption, or both. Passenger trains, as noted earlier, may be the targets themselves if they are intercity. While commuter trains may also be targeted, the commuter lines have in addition greater exposure at the many stations served. The freight railroads are a twofold target: both by disrupting economic activity through severing critical transportation lines needed for inbound raw materials and thereby interrupting manufacturing activities, or causing death and destruction by targeting those trains hauling hazardous materials, or using them to convey explosive, biological, or even nuclear materials.

The infrastructure of interest to terrorists a mere half century ago would have been the rights-of-way and associated track, bridges, and tunnels. However, nowadays one must also consider the communication and signaling systems and the various freight and passenger terminals that we earlier explained as key nodes of the network.

The task of protection is confounded by the number of legal jurisdictions involved, both vertically and horizontally. In the case of the former, there are the layers of law enforcement at the federal, state, tribal, and local levels to which is added the law enforcement roles that over a century ago were ceded by the various state legislatures to the railroads. Horizontally, there are the thousands of local governments with their varying amounts of police and other first responder resources: mosaics of differing, and in some instances contradictory, regulatory regimen.

Countervailing assets

Citing some previous work published by the Transportation Research Board, the four objectives for protecting transportation infrastructure in priority order are: deter, detect, deny, and mitigate (TRB 2006, p. 9). There is a range of assets that can be applied to these objectives. The following matrix suggests that there are combinations of factors that may be brought to bear with good effect. Table 11.1 juxtaposes the potential security objectives with the potentially impacted asset type.

Table 11.1 Objectives and asset type

	Deter	*Detect*	*Deny*	*Mitigate*
Human assets	• Active surveillance to frustrate targeting and attack planning	• Police patrols, but also an aware workforce and the public-at-large	• Perimeter and access control via patrols	• Design human assets to maximize presence and visibility
Technology applications	• Perimeter fencing and surveillance	• Alarms, motion detectors, and cameras	• Scanners, codes access cards and physical barriers	
Network modification	• Barriers, fencing, and rerouting to avoid		• Relocate high value or target assets	• Modify rail cars, stations, and terminals to lessen the effect of attacks
Intra-industry combinations		• Intelligence gathering: rail industry fusion centers, but also AAR and ASLRRA efforts		• Improve rail car design standards to minimize attack effects
Inter-industry combinations		• Intelligence gathering: customer trade associations such as American Chemical Council		• Improve the availability of actionable information about handling hazmat
Public policy initiatives	• Pass regulations for more severe penalties for trespassers	• Provide more funding for research, but also increase budgets for policing, especially for public-private cooperation	• Provide more funding for research into new technologies	• Reduce legal liability of railroads for incidents with trespassers

Source: compiled by the authors

Human assets

In the past the railroads looked at human assets for protecting infrastructure as a matter of employing their own police forces to protect company property, passengers and their personal property, and cargo. For the most part policing was focused on the actions of individuals or small and somewhat undisciplined groups of individuals. While lone terrorists remain a threat, the new reality is that there is likely to be an orchestrated attack as demonstrated by the synchronized bombings in Madrid and London. In short, the protection effort required exceeds the capabilities and capacities of the typical railroad police agency. The way forward is paved with cooperation, but cooperation between railroads; with first responder agencies, both public and private; between the railroads and the public; and among new organizations that we prefer to think of as cross-industry combinations. A best-case example of a cross-industry initiative would be the chemical industry and the railroads working to improve safety when transporting hazmat.

It is appropriate to reiterate that railroad police agencies continue to evolve rapidly and can no longer be thought to be reactive crime-solving forces, but instead they are becoming proactive intelligence gathering professionals armed with the same modern technology and other tools as their public sector counterparts. Some we have noted, such as with BNSF, have their own SWAT teams and K-9 units and actively participate in fusion centers.

Intelligence units, many still in their infancy, can be expected to become more widespread, if not commonplace in the future.

Technology applications

There are some basic technologies extant that can be employed to better protect railroad infrastructure, including, but not limited to, those falling into the categories of surveillance and entry restriction (aka access control). Cameras and intrusion sensors, in addition to alert personnel, are useful for watching for the presence of individuals or groups either entering upon railroad property or attempting to engage in an act against infrastructure. Fencing and appropriate lighting, while neither are high-technology solutions, are useful for limiting access or, in the case of the latter, preventing unobserved egress.

Aircraft are useful for surveillance of large expanses of remote areas; however, as the pipeline companies are discovering, the use of drone aircraft may be more cost effective and, hence, provide greater coverage. As this chapter is being written, the western U.S. railroads have made the news with their use of drones for patrolling large expanses of open space along their rights of way. With the prices decreasing rapidly for this technology, it easily becomes affordable for even the modest budgets for many of the nation's shortlines.

It was noted that train control systems for both Amtrak and the Class I's are now computer controlled, thus leaving train dispatching and signal systems open to cyber terrorism attack.[1] Some systems are more prone to hacking than others, with some controlled by nothing more sophisticated than a desktop personal computer. The ability to hack into systems that control track switches or signal aspects has the potential of causing derailments and shutting down major lines with little chance of detection until it might be too late to avert catastrophe. For some commuter lines and Amtrak, the ability to curtail power to those electrified lines that run off third rail or overhead catenary is a very real threat. Many substations are located in isolated locations and/or provide quite little in the way of physical protection. Much has been written about the ease of attacking the power grid in the U.S., but little mention is made of the vulnerability of electrified railroad lines, which in some respects are nothing more than a microcosm of a regional grid.

Network modification

Removing or relocating stations, track, or other vulnerable infrastructure is perhaps one of the most expensive alternatives that can be considered to reduce vulnerability. Risk assessment techniques may be employed for identifying those locations that may be the most problematic. In some cases for commuter railroads, this could be as simple as the elimination of stations where the locations are remote and the ridership negligible. These stations cannot be adequately policed, but nevertheless they represent a potential point of entry to persons wishing to gain access to the system unobserved or to conduct an attack along the route or at the terminal station. Rerouting hazmat trains, while not a modification of the network, actually uses the network's redundancy to shift traffic to lower-risk venues – FRA's Rail Corridor Risk Management System (RCRMS) is particularly noteworthy in this regard.

Public policy initiatives

As discussed earlier, the development of rail security policy since the 9/11 attacks has been reactive to events, incremental in development, and based on collaborative arrangements between TSA, state, tribal and local governments, and rail operators. The goal remains to develop a proactive approach, based on an understanding of the nature of threats to the rail system and its vulnerabilities. Programs founded on risk-based analysis have, since the mid-2000s, moved rail security from a virtual stepchild of homeland security policy to a more professionally grounded area of expertise in which information sharing, training exercises, initiatives to protect infrastructure, and traditional policing and surveillance prevail.

Looking back over the development of rail security policy, it seems that four approaches to developing policy have been employed. Legislation, beginning with the PATRIOT Act and moving through various subsequent acts devoted to transportation security, was a necessary foundation for the creation of TSA and DHS, and in identifying priorities for the nation's approach to homeland security. Passage of the Implementing the 9/11 Commission Act of 2007 was a major step forward in bringing together the insights learned in the formative years after 2001 and in specifying priorities for DHS and the nation to pursue. In recent years, the partisan deadlock in Congress, and the absence of compelling catastrophic rail attacks such as the Madrid and London bombings of the mid-2000s, have left rail security a topic that is not likely to move forward through legislation, pending major external events forcing the hand of Congress.

A second major approach to homeland security policy has been the use of presidential directives: top-down executive orders to guide actions within and among the executive branch agencies. Presidential Policy Directives (PPD) were fundamental in the early years of DHS to add flesh to the bones of policy created by the PATRIOT Act and legislation creating the department by charging DHS and other participating federal agencies to move in specific directions. The most current such effort is PPD 8 issued by President Barack Obama on March 30, 2011. This directive charges the secretary of DHS to coordinate planning for homeland security and counterterrorism and produce a set of National Preparedness Goals and a methodology for measuring operational readiness.

A third approach (and one that is included in the most recent presidential directive) has been the development of collaborative efforts with the other stakeholders involved in rail security: state and local and tribal governments, their law enforcement agencies, rail operators and industry representatives, and various federal agencies. The goal of such arrangements has been to develop a common vocabulary around which to share information, better understand the nature of threats and vulnerabilities, plan for contingencies, and break down barriers by recognizing and accepting the roles each can and must play in securing the nation's rail network. At the center of the network by law and executive order is the TSA, but given resource limitations and the necessary priorities for its top leadership to be airline security, it must rely on network partners to be ready to do the heavy lifting required to prevent attacks and respond as needed to catastrophes.

A fourth and final way policy is made and implemented is by executive branch action that is bottom-up and horizontal rather than through top-down directives. Since roughly 2005 an influx of talented and experienced rail experts transformed TSA from an agency concerned almost entirely with airline security to one that could speak with authority on the topic of rail security. Since then, TSA has been able to use its rail expertise to guide incremental change in the programs and activities directed toward securing the rails. Resource limitations and the lack of individuals able to champion the cause of rail security in Congress remain obstacles. Nevertheless, the record in the past decade is that TSA leadership, although constrained by the need to devote most resources to screening air passengers and cargo, has recognized and supported activities needed to ensure that surface modes such as rail and maritime transportation have at least begun to provide the minimal personnel and resources needed to expand the understanding of risk and, most importantly, have the expertise and experience needed to be seen as legitimate and valued members of the rail security network. Nonetheless, TSA's focus remains largely on domestic and international airline travel, with rail security well down the list of priorities.

In short, public policy regarding rail security is still heavily responsive to external events, both in the U.S. and in other nations, but it is informed by an increasingly solid foundation in shared values, assumptions, and years of experience sharing information and working to prevent attacks.

Other combinations – intra-industry

As we previously established, there are two primary organizations that have been in place for many years whereby railroads can establish common standards whereby operations between them can function with a degree of seamlessness, specifically the Association of American Railroads (AAR) and the American Short Line and Regional Railroad Association (ASLRRA). As organizations between competitors, these have long sought to support best practices, whether they apply to the specifications for equipment used in interchange, standards for tank cars intended to carry hazmat (as discussed in earlier chapters), or the exchange of security-related information between members.

Other combinations – cross-industry

The protection of railroad infrastructure turns out to be an effort that demands the participation of all potential stakeholders. In an earlier chapter, we made passing note of the long-standing efforts of the American Chemical Council (now ACC, but formerly the Chemical Manufacturers' Association), which has had working groups focused on terrorist threats and the potential use of chemicals for illicit purposes, such as refining recreational drugs, dating back into the 1980s. Moreover, their creation of ChemTrec was an early initiative to mitigate the impact of any hazmat event. From the beginning the railroads have had an ongoing relationship with the ACC.

Public-private partnerships

Previous mention has been made of fusion centers, which are formal organizations stood up to exchange threat information. Railroads have a clear history for participating with these, but on a more ad hoc basis the railroads have a history of holding joint exercises with state and local first-responder organizations nationwide. However, the term *ad hoc* needs to not only be repeated but also emphasized because there are nearly as many examples as there are permutations of freight railroad companies, passenger operating agencies, and political jurisdictions. After conducting the research for this book, we have come away holding the view that the public-private partnerships need to be more steadfastly encouraged nationwide. However, given the differences in geography, locations of population centers, and what has been termed venues of interest, a decentralized approach has many attractive attributes, not the least of which is personal relationships, knowledge of the local environment, and faster response.

In summary

In short, there is no single "magic bullet" that can be employed to protect the nation's trains and rail infrastructure. First, the geographic diversity of the rail system and the number of organizations involved in rail operations permit far too many permutations for any one approach to ever be able to fit all. Second, the idea that there is a 100 percent solution to the problem is an unrealistic one, and, even if it did exist, it would not likely be economically feasible to deploy. Third, the U.S., as a free and open society, cannot put substantial limitations on its citizenry such that their movements are unduly constrained. Fourth, acceptable protection, and we note that this cannot be total protection, will be achieved only through a comprehensive and holistic effort whereby individual rail operators, a wide array of government agencies with an almost equally wide range of operating objectives, and the public at large are engaged. Fifth, there needs to be a recognition that both threats and the means of executing such threats shall continue to evolve – hence, those approaches employed for thwarting such threats need to continue to evolve as well. Given that we cannot obtain perfect information about those future threats, intelligence gathering that allows the railroads to be fighting the next and future threats need to be seriously considered. Here an emphasis is intentionally placed on those threats categorized as cyber attacks or cyber terrorism, but this does not mean as a replacement for traditional efforts to protect infrastructure – this only implies that terrorists will have a larger menu of options from which to select.

The protection of infrastructure is something that is akin to a race without a clear end in sight. However, given the real vulnerabilities of the vast U.S. rail network, it is a race that still must be run. Protecting railroad infrastructure constitutes the challenge of protecting the populace from harm, while also protecting the economic wellbeing of the nation. Finally, we must reiterate that protecting railroad infrastructure, while there is much concern about the potential of terrorist threats, still requires a concern for naturally caused threats, which in recent years have occurred with far greater frequency and in the aggregate have incurred far greater damage. And what about the exploitation of a natural disaster by a terrorist? The all-hazards approach to security is critical to any approach to rail security.

The book in retrospect

This has been a challenging book to write for several reasons. First, the U.S. rail system is complex, consisting of hundreds of companies, commuter rail operators, and Amtrak. There is lots of overlap of rights-of-way, regulation, and police agencies. If there is an Achilles heel, it is probably found in three specific areas: (1) the ongoing need for railroads to form collaborative partnerships with other types of organizations that must be involved in rail security; (2) the lack of public policy emphasis on rail transportation, stemming from airlines having the highest profile in the eyes of the public and, thus, the major share of federal initiatives and budgeted resources within the overall homeland security arena; and (3) the inherent characteristic of a system that is an expansive and virtually unprotected nationwide network.

Much of the perceptual problem is based on the past, when railroads were the major mode of transportation and made subject to the first attempts at economic regulation. Even now, railroads are still trying to get over their history of fear of being regulated again and are still always comfortable working with government and the public. However, since all modes of transportation were economically deregulated beginning in the late 1970s and early 1980s, the "us vs. them" mentality of the past has receded. Nonetheless, ask the average citizen about the condition of the railroads, and many will insist that the railroads are an antiquated and obsolete mode of transportation. We have tried to argue how antiquated that position is by illustrating the huge portion of basic materials and intermodal traffic moves that the country depends upon for its economic vitality. Passenger rail receives much publicity due to Amtrak's ongoing financial issues with Congress, but the highly successful freight railroads, clearly the envy of the world in their ability to haul enormous tonnage economically, safely, and in a timely manner, still take a backseat.

The network remains vulnerable to both natural and manmade disasters or an exploitation of the latter by groups and individuals desiring harm to the U.S. Vigilance needs to be maintained by the individual companies and operating authorities, their industry associations, government at all levels, and the public at large. It is a daunting task, and our parting hope is that we have raised the consciousness through an elevation of the discussion by reminding everyone involved of the myriad factors that need to collectively come into play to ensure that the nation's freight and passenger rail system remains a safe, secure, and vital part of our transportation industry.

Note

1 There is ample coverage in both the mass media and the industry press about the threats posed by employee, the use of thumb drives, and viruses introduced by connectivity to the Internet, whether through emails or downloads from websites. With such coverage plus its rapid development, the decision was made to exclude this from this book.

Bibliography

Transportation Research Board (2006). Security Measures for Ferry Systems. *TCRP Report 86 Public Transportation Security*, Vol. 11. Washington, DC: Transportation Research Board of the National Archives.

Index